THE FIRST WORLD WAR

THE FIRST
WORLD WAR

Robin Prior and Trevor Wilson

General Editor: John Keegan

CASSELL&CO

For Grahame
R.P.

For Lucy, Harriet and small Ben, top grandchildren;
and for Nirej, Mark and Tim, companions in table tennis.
T.W.

Cassell & Co
Wellington House, 125 Strand
London WC2R 0BB

First published 1999
Reprinted 2000

British Library Cataloguing-in-Publication Data
A catalogue record for this book is available from the
British Library.
ISBN 0-304-35256-X

Cartography: Arcadia Editions
Picture research: Elaine Willis
Design: Paul Cooper Design
Printed and bound in Italy by L.E.G.O. S.p.A.

Typeset in Monotype Sabon

ACKNOWLEDGEMENTS

The authors would like to acknowledge the assistance and encouragement they have received from their colleagues in the History Departments at University College, Australian Defence Academy, and the University of Adelaide. In preparing the typescript for publication Elizabeth Greenhalgh has been, as always, indispensable. In this area many thanks are also due to Robyn Grccn, Julie McMahon and Julie Cassell. At our publishers, our editor Penny Gardiner has applied her expertise, patience, and wisdom to produce many improvements to the text and format of the book. No editor could have been more pleasant to work for.

Finally, and above all, the authors wish to thank their wives, Heather and Jane, for enduring another book on the First World War with their customary critical interest and good humour.

ROBIN PRIOR
TREVOR WILSON

Forward post in a German sap on the Galician front, 1916.

CONTENTS

KEY TO MAPS

Military units – types

⊠ infantry

▬ armoured

▱ motorized infantry

⌒ airborne

☂ parachute

● artillery

Military units – size

XXXXX
▢ army group

XXXX
▢ army

XXX
▢ corps

XX
▢ division

X
▢ brigade

III
▢ regiment

II
▢ battalion

Military unit colours

▬ Allied

▬ British

▬ German

▬ French

▬ Austro-Hungarian

▬ Russian

▬ Italian

General military symbols

—XXXXX— army group boundary

—XXXX— army boundary

⌣ front line

ᗡᗞ defensive line

⊓⊔ defensive line (3D maps)

⌄⌄⌄ field work

◯ pocket or position

⊢ field gun

☂ paratroop drop

⨺ sunken ship

✈ airfield

Geographical symbols

▱ urban area

—— road

═▬═ railway

—— river

- - - seasonal river

⊥⊥⊥ canal

—— border

⊃⊂ bridge or pass

Military movements

➤ attack

- ➤ retreat

✈ air attack

MAP LIST

CHRONOLOGY

1914

June 28 The heir to the throne of Austria–Hungary, Archduke Franz Ferdinand, is assassinated by Bosnian nationalists at Sarajevo (then part of the Austrian empire).

July 6 Germany gives assurances of support to Austria–Hungary should its rulers decide to take action against Serbia.

July 28 Austria–Hungary declares war on Serbia.

July 30 Russia mobilizes in support of Serbia.

July 31 German ultimatum to Russia.

Aug 1 Germany declares war on Russia.

Aug 2 German ultimatum to Belgium.

Aug 3 Germany invades Belgium, France and Luxembourg in accordance with the Schlieffen Plan.

Aug 4 Britain declares war on Germany in response to Germany's refusal to withdraw from Belgium.

Aug 7 Belgian fortress of Liège captured by Germans.

Aug 11–25 French army invades Alsace–Lorraine and is repulsed with huge losses.

Aug 16 British Expeditionary Force completes disembarkation in France.

Aug 20 German troops enter Brussels.

Aug 23–30 German armies on the right of their line drive French–British forces all the way to the River Marne.

Aug 25 Austria–Hungary launches campaign in Galicia.

Aug 26 French fortress of Maubeuge surrounded.

Aug 26 German forces in Togoland surrender.

Aug 26–31 Russian northern army invades East Prussia and is defeated at Tannenberg.

Aug 27 Germans capture Lille.

Sept 3 Russians counter-attack in Galicia.

Sept 5–10 Germans defeated at the Battle of the Marne.

Sept 8 Austrians invade Serbia for the second time.

Sept 9–14 Russians defeated at the Masurian Lakes and driven off German soil.

Sept 15 Austrians driven out of Serbia.

Sept 23 Siege of Tsingtao by Japanese begins.

Oct 10 Antwerp surrenders to the Germans.

Oct 29 Turkey enters the war on the German side.

Oct–Nov First Battle of Ypres, as Germans in the west make unsuccessful thrust for the Channel ports.

Nov 5 British forces land at Basra in Mesopotamia.

Nov 7 Japanese capture Tsingtao.

Dec 20 French attack inconclusively in Champagne but persist in maintaining the offensive until March 1915.

1915

Jan–Feb Carpathian campaigns by Austro-Hungarians and Russians end indecisively.

Jan 26 Turks repulsed in attack on Suez Canal.

Feb 19	Naval attack on the Dardanelles begins.	Sept 25	Allied offensive in Artois and in Champagne.	April 10	Germans commence attack on the left bank of the Meuse at Verdun.
March 10	Small-scale British success at Neuve Chapelle.		British forces in Mesopotamia capture Kut.	April 29	British forces at Kut, besieged by the Turks for several months, surrender.
March 22	Austrian fortress of Przemysl in Galicia surrenders to Russians.	Oct 6	Germany, Austria–Hungary and Bulgaria invade Serbia	May 14	Austrian offensive in the Trentino.
April 22	Germans launch gas attack at Ypres.	Oct 7	Belgrade captured by Austrians.	June 4	Commencement of Brusilov offensive against Austro-Hungarians.
April 25	Allies land in Gallipoli to facilitate projected British naval attack on Constantinople.	Oct 8	End of the British action at Loos.	June 5	Lord Kitchener drowns when his ship is torpedoed en route to Russia.
		Oct 9	British and French forces land at Salonika.		
May 2	German offensive breaks through at Gorlice–Tarnow in Galicia.	Nov 16	End of French action in the Champagne.	June 18	Last German forces in Cameroon surrender.
May 8	Nicaragua declares war on Germany.	Nov 23	Allies decide to evacuate Gallipoli peninsula.	June 24	Limit of German advance at Verdun.
May 9	French attack in Artois.	Nov 27	Defeated Serbian army evacuated to Corfu.	July 1	Opening of Anglo-French campaign on the Somme.
May 23	Italy enters the war on Allied side.	Dec 19	Sir John French is replaced as British commander-in-chief in the west by Sir Douglas Haig.	July 14	Major British attack on the Somme captures sections of German main defensive line.
June 3	Przemsyl captured by the Central Powers.				
June 16	End of the French offensive in Artois.				
June 23	First battle on the Isonzo between Italians and Austro-Hungarians begins.	**1916**		Aug 27	Romania enters the war on Allies' side.
July 9	Surrender of German forces in South-West Africa.	Jan 8–9	British evacuate Helles, ending the Gallipoli campaign .	Aug 28	General Maude takes command in Mesopotamia.
Aug 5	Germans enter Warsaw.	Feb 21	Germans attack French fortress system at Verdun.	Aug 29	Hindenburg and Ludendorff assume command in Germany.
Aug 6	British launch last major attempt to overrun Gallipoli Peninsula, without success.	Feb 26	Germans capture Fort Douaumont, a key fortress of Verdun's defences.	Sept 15	Tanks used for the first time in warfare in British operation on the Somme.
		March 18	Russian offensive at Lake Naroch.		

Oct 24 — French counter-attack at Verdun.

Nov 18 — British operations on the Somme end.

Nov 19 — Allied forces capture Monastir in Balkan offensive.

Dec 6 — Romanian capital, Bucharest, captured by Austro-German forces.

Dec 7 — Lloyd George becomes British Prime Minister.

1917

Jan 31 — Germany announces unrestricted submarine warfare.

Feb 23 — German forces in the West commence withdrawal to Hindenburg Line.

Feb 24 — Kut recaptured by British forces.

March 11 — British enter Baghdad.

March 12 — First Russian Revolution.

March 14 — German retreat to the Hindenburg Line begins.

March 15 — Tsar abdicates.

April 6 — United States declares war against Germany.

April 9 — Opening of British offensive at Arras enjoys brief success, including the capture of Vimy Ridge.

April 15 — Battle of Arras ends.

April 16–20 — Nivelle offensive against the Germans in Champagne reduces French army to mutiny (29 April).

May 15 — Pétain becomes French commander-in-chief.

June 7 — British attack in the Ypres salient captures Messines Ridge.

June 18 — Kerensky offensive.

June 29 — General Allenby assumes command of British forces in Palestine.

July 31 — Third Battle of Ypres commences.

Sept 3 — Germans capture Riga.

Sept 20–Oct 4 — Three successful limited objective operations conducted by General Plumer in the Third Ypres campaign.

Oct 24 — Italian defeat at Caporetto.

Oct 27 — Allenby commences battle for Gaza.

Oct 29 — Austro–Germans continue advance from Caporetto and capture Udine.

Nov 7 — Bolshevik seizure of power in Russia.

Nov 7 — British advance in Palestine.

Nov 9 — General Diaz takes over command of the Italian army from General Cadorna.

Nov 10 — Third Ypres campaign ends.

Nov 19 — Clemenceau becomes French premier.

Nov 20 — Battle of Cambrai.

Dec 9 — Allenby captures Jerusalem.

Dec 10 — Armistice between Central Powers and Romania.

1918

Jan 8 — President Wilson issues 14 Points.

March 3 — Treaty of Brest-Litovsk imposed by Germany on Russia.

March 21 — Ludendorff offensive against British in Somme region.

March 26 — Doullens conference appoints Foch to command of French and British armies on the Western Front.

March 27 — German troops enter Ukraine.

April 9 — Germans attack against British in Flanders.

April 12 — Haig issues 'Backs to the Wall' message. German troops occupy Helsinki.

April 19 — German troops enter Crimea.

April 24 — Germans resume advance on Amiens.

April 25 — British and Australian troops halt German advance on Amiens at Villers-Bretonneux.

May 26	Georgia and Armenia declare independence from Russia.	Sept 14–15	Baku attacked by the Germans.	Nov 8	Maubeuge retaken by the British.
May 27	Germans attack on the Aisne.	Sept 18	British commence operations against the Hindenburg position.	Nov 9	Abdication of the kaiser.
June 9	Germans attack at Noyon.	Sept 19	Last major action in Palestine commenced by Allenby.	Nov 11	Armistice signed with Germany.
June 15	Italians halt Austrians along the Piave.	Sept 25	Bulgaria seeks an armistice.	**1919**	
June 18	General Franchet d'Esperey appointed commander-in-chief allied forces at Salonika.	Sept 26	Combined French–American offensive in the Meuse–Argonne.	Jan 4	Peace conference begins at Paris.
				June 28	Treaty of Versailles signed.
July 15	Last German offensive on the Western Front begins near Reims.	Sept 27–Oct 5	British attack; breach Hindenburg Line.	Nov 19	US Senate refuses to ratify Treaty of Versailles.
		Sept 28	Ludendorff informs the Kaiser that an armistice should be sought.		
July 16	Conrad sacked as commander of Austro–Hungarian forces.				
		Oct 1	Damascus falls to the British.		
July 18	French counter-attack on the Marne.	Oct 6	German government makes first attempt to negotiate an armistice.		
Aug 4	Small British force occupies Baku in the Caucasus.	Oct 8–9	Cambrai captured by the British.		
Aug 8–12	British counter-attack at Amiens.	Oct 21	Czechoslovakia declares independence.		
Aug 18	British advance in Flanders begins.				
Aug 21	British advance across old Somme battlefield begins.	Oct 24	Italian forces commence battle of Vittorio–Veneto.		
Aug 29	British recapture Bapaume.	Oct 26	Ludendorff replaced.		
		Oct 30	Turkey concludes an armistice.		
Sept 12	Americans launch offensive at St Mihiel.	Nov 4	Armistice concluded with Austria–Hungary.		
Sept 14–15	Allies attack Bulgarians at Salonika.	Nov 4	Final Allied offensive launched on Western Front.		

THE COMING OF WAR

*THE APPLICATION OF INDUSTRIAL TECHNOLOGY
to the weaponry of war, in a time of international
tension, witnessed the development of huge
armaments manufacturers. Shown here is the
Krupp factory in Germany in 1909.*

THE COMING OF WAR

FIRST THOUGHTS

Few events of the past so tax the historian's powers of explanation as the onset of great wars. This is not, usually, because lines of explanation are lacking. It is because the explanations on offer are so many and varied – not to say contradictory and mutually exclusive.

At one time (by way of an example that is not the subject of this book), the outbreak of the Second World War seemed easy to account for. It appeared the self-evident consequence of the unique malevolence and aggressive urges of a single ruler (Adolf Hitler), a single governing force (the Nazi party) and a single nation (Germany).

Then, with the appearance in 1961 of A. J. P. Taylor's *The Origins of the Second World War*, such simplicities vanished. Taylor, with characteristic exuberance, incorporated in his book a startling variety of explanations as to why a European war and then a world war broke out between 1939 and 1941. The assertion which most seized attention, and most excited controversy, was Taylor's argument that the Second World War did not spring from the deliberate intent or lust for conquest of any power or person. It was the result of an accidental conjunction of fortuitous events.

At the time he wrote, it may be noted, Taylor was an advocate of the Campaign for Nuclear Disarmament. In that capacity he was seeking to refute the widely held view that the possession of nuclear weapons acted as a deterrent to potential aggressors and so prevented wars. Taylor was concerned to demonstrate the contrary: that great wars are caused not by intent or well-established plans, but by chance, accident and miscalculation. So, he argued, the Second World War broke out not because any nation intended it but because governments were engaged in familiar games of bluff and brinkmanship. Their intention was to secure advantages at the expense of their neighbours without actually generating conflict. On this occasion, however, the process went wrong. To quote Taylor's riveting judgement, war erupted in September 1939 because Adolf Hitler 'launched on 29 August a diplomatic manoeuvre which he ought to have launched on 28 August'.

Curiously, this is not the only level of explanation offered in *The Origins of the Second World War*. Taylor also locates the onset of that conflict well back in history: in the long-standing animosity between France and Germany, and not least in the ambiguous and unresolved outcome of the Great War of 1914–18 – an outcome encapsulated in the plainly unsustainable Treaty of Versailles of 1919. At the same time, Taylor apparently reverts to a judgement he had once held forcefully but seemed now to be abandoning. He proclaims that the Second World War – certainly as a *world* war – was solely the result of Hitler's and Germany's unprovoked and gratuitous acts of aggression. In this third view,

Hitler without cause unleashed his forces against 'two World Powers' (the Soviet Union and the USA) 'who asked only to be left alone'.

The point might be made that these lines of argument, being at best inconsistent and for the most part flatly contradictory, scarcely belong in a single work of historical explanation. But that was not, overall, the response to *The Origins of the Second World War*. Rather, it has been assumed that an event as cataclysmic as a world conflict requires explanations that are multi-faceted, located on different planes, and even mutually exclusive. No doubt this willingness to entertain what appeared to be Taylor's leap into complexity, uncertainty, and – dare one say it? – downright nonsense had been well prepared for. It followed from the bewildering range of explanations which had come forth to account for the onset of a previous great conflict. That conflict was the Great War of 1914–18, which does happen to be the subject of this book.

THE VARIETY OF HISTORICAL EXPLANATION

From the moment of its outbreak, dispute has attended the matter of responsibility for the coming of war in 1914. As against the simplistic views proclaimed by the warring governments, namely that battle was the consequence of aggression by one power or group of powers, quite different explanations were soon being offered.

Early in the war, a Russian political exile living in Switzerland delivered the judgement that the international explosion was the product of large economic forces embodied in the capitalist system. In his view, the militant imperialism which capitalism, by its very nature, had generated led inexorably to this bloodbath.

At a quite different level, the view was strongly asserted, again while the conflict raged, that the war was the responsibility not of peoples or of economic forces but of the 'old diplomacy'. That is, a clique of unelected permanent officials in the foreign offices of the various powers had, wilfully or unintentionally, worked great evil. They had bound the countries of Europe into a system of competing alliances that left no freedom of choice or room for manoeuvre when a particular quarrel erupted in one remote region. Where otherwise such a dispute could have been contained, the alliance obligations of the powers dragged first one nation and then another into a Europe-wide maelstrom.

Other explanations soon emerged and gained endorsement. War was seen as the irresistible product of fears and enmities generated by changes in the established power balance within Europe. From 1870 Germany had grown in population and economic strength to the disadvantage – and resentment – of its neighbours. And during the same period nationalistic Balkan states, often patronized by Russia, had shaken off the shackles of Turkey and were threatening the multi-national basis of the Austro-Hungarian empire.

Another view locates the source of the Great War in the internal instability of established regimes threatened by powerful discontent at home. In this view, ruling élites whose hold on power was becoming contested by the aspirations of an alienated proletariat or militant feminists or ethnic minorities chose to provoke external conflict as a means of allaying pressures within. More recently, this type of argument has transformed itself into another, which explains international violence as the emanation of a profound sense of disempowerment among the masses within advanced industrial societies. That is, the dispossessed of one nation, instead of directing their wrath against the exploiting classes within their own countries, developed a paranoid antipathy towards imagined enemies beyond their borders. This rendered them responsive when their national leaders summoned them to arms.

Such wide-ranging, 'in-depth' explanations have not won universal endorsement. Reference may again be made to A. J. P. Taylor in his CND phase. Drastically revising his earlier (*Course of German History*) views about the causes of the Great War – as he was doing simultaneously for the causes of the Second World War – he concluded that large events such as those of July–August 1914 need not have been the product of large antecedents. They may have sprung from a succession of trivialities: chance, accident, diplomatic manoeuvres that went awry, declarations of war that were intended as bluff rather than as a determination to prosecute conflict, and plans for war mobilization that placed military movements outside human dictates and at the beck and call of railway timetables. Europe, Taylor proclaimed, was as peaceful in 1914 as at any time in the preceding half-century. But fortuitous events and diplomatic misjudgements took over, deterrents failed to deter, and mobilization schemes designed to preserve the integrity of nations swept Europe into self-destruction.

MAKING A CHOICE

How do we deal with this bewildering range of proffered explanations for the outbreak of the First World War? One thing is not possible. These varied accounts cannot be merged into a single all-encompassing explanation. If we see the war as the product of deliberate predatory intent on the part of the rulers of one great power, then chance and accident and developments in transportation are not significant. If we regard the war as springing from the imperatives of capitalism in its most advanced state, then it is pointless to invoke the ambitions and fears of less-developed countries like Russia and Austria–Hungary, or the instability generated by developments in the Balkans. If we wish to view the war as the product of a great psychological upheaval undergone by the masses of western Europe passing through the agonies of industrialization, then it hardly matters what the diplomats were up to or what agreements they had surreptitiously entered into. Somehow, forms of explanation must be assessed so as to eliminate those with insufficient substance and to test the potency of those that remain.

One thing had better be said at the outset. War occurs because the great mass of human beings are prepared, at least in certain circumstances, to regard the resort to arms as an acceptable proceeding. They may wish to enrich their communities, and so enhance their own self-esteem, by engaging in predatory acts at the expense of their neighbours. Or they may only be prepared to engage in battle to resist what appears to be the aggression and violence of others. Either way, a deliberate choice for war is being made. In that respect, even 'little Belgium' helped to cause war in 1914. If Belgium had lain down and allowed the German army to overrun its territory, the consequences for the Belgian people could well have been profoundly unpleasant. But the clash of armies would not have been among them.

Such generalities remind us that war is located in human nature. But they do not answer the particular questions. In the circumstances of 1914, were some nations engaging in acts of aggression, and if so under what compulsions, while others were seeking only to preserve their territorial integrity against attack (an attack which they may, or may not, have done something to provoke)? Or were all nations, or anyway the principal nations on both sides, contributors to a state of international disharmony so intense that it boiled over – perhaps for trivial reasons – into active conflict?

As a first step in resolving these matters, it is appropriate to set aside as

Although carefully keeping itself outside the system of rival alliances, Belgium did not manage to escape involvement in the conflict. German troops enter Antwerp on 9 October 1914.

Einzug der deutschen Truppen in Antwerpen am 9. Oktober 1914

downright unsatisfactory some of the explanations for the coming of war which have been referred to. One among them is the Leninist view which regards capitalist imperialism, in the form of rivalry in the colonial field conducted by state instrumentalities for the benefit of the possessing classes, as the pre-eminent force generating international conflict in 1914. What Lenin was accounting for was conflict between Britain and Germany, the two most industrialized powers, which he postulated were bound to come to blows on account of their increasingly desperate struggle for markets and raw materials. Nothing about the actual course of events bears this out. Plainly Britain's colonial rivalry with France and Russia, neither of which was anywhere near the highest stage of

EUROPE'S IMPERIAL
POSSESSIONS IN 1914

The world-wide distribution of European colonies and often their proximity led to the conflict spreading to such unlikely areas as the Pacific, Africa and the Middle East. So, because of these minor skirmishes, what was essentially a European war became known as a world war.

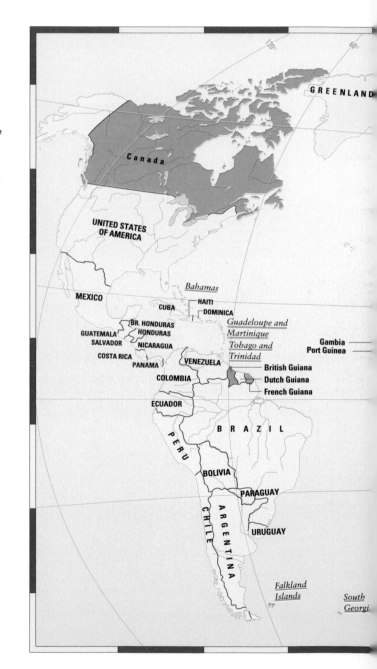

capitalism, had been at least as intense as its overseas competition with Germany. Indeed, the only formal alliance that Britain entered into before the First World War was with Japan, quite manifestly directed against Russian expansionism.

When Britain in the early twentieth century underwent a diplomatic revolution, by which it established at least friendly understandings with France and then Russia, thereby placing on hold its colonial animosities towards them, this was not on account of anything that was occurring in the colonial field. It was in response to what was threatening to happen on the continent of Europe and in the North Sea, areas in which Britain's vital interests were centred to an extent that never applied to its possessions overseas.

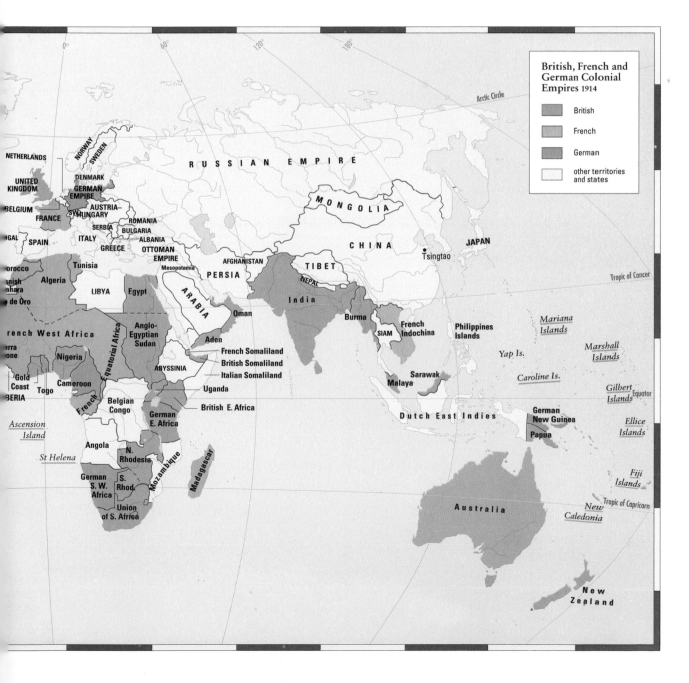

British, French and German Colonial Empires 1914

- British
- French
- German
- other territories and states

The rival alliances (1): Franz Ferdinand, heir to the throne of Austria-Hungary, with Wilhelm II of Germany at Kiel, a major naval base of the German battlefleet. The creation of this fleet drove Britain into association with France and Russia.

Equally unsatisfying, for all its attractiveness, is the view put forward by A. J. P. Taylor that the Great War sprang only from accident and miscalculation, not from any more profound causes. The existence of deep-lying forces generating antagonism between states had manifested itself often enough. Those forces had been gathering strength at least since the emergence of a united Germany. On top of its stunning defeat of France in 1870–71, Germany by the start of the twentieth century was developing its economic and military power to a huge extent and was also indulging aspirations to big-navy status. These developments perforce constituted a threat not just to the position of Britain and France as major powers but to their very independence. This source of tension was not lessened by the persistence into the twentieth century of aspects of Bismarckian diplomacy which involved Germany in the threatened internal instability and drives to external action of its ally Austria–Hungary (and so in the potential rivalry of Austria–Hungary with tsarist Russia). Chance events, such as the assassination of an Austrian arch-duke and even the development of railways as a means to transport armies, only mattered in a world where the urge to act on larger animosities continued to be powerful.

As for the notion that war was caused by the mounting alienation and psychological disturbance of the masses consequent upon the vagaries of the trade cycle and the sense of powerlessness engendered by industrialization, this again bears no apparent relation to the actual course of events. If the consequence of industrialization had been to transform the formerly settled masses into a vast body of alienated, déraciné and so potentially aggressive people, it would follow that the most provocative, armed and war-eager states in 1914 would be Britain, the USA and Belgium. Germany would fall well behind in such inclinations, with France even further back and Russia and Austria–Hungary not at all disposed towards conflict. Nothing about the actual sequence of events, either in the long term or in the immediate crisis of July–August 1914, sustains such an interpretation.

A different sort of objection must be raised to the widely held view that war was imposed on the people of Europe by the 'old diplomacy' and the system of 'binding alliances' which it supposedly generated (in Britain's case not even alliances, only 'moral obligations'). Nations, it needs to be pointed out, enter into alliances and understandings because they see these as serving their vital interests: enabling the participants to secure mutual advantages or hold at bay mutual threats. If the arrangements do not promise to fulfil these purposes, or if ambitions and perils undergo change, then the alliances fall into abeyance or are simply not acted upon. In this respect, we should pay attention to the point often made that there seemed a distinct possibility in the 1890s that Britain, on account of its antagonism to France and Russia, might enter into an alliance with the Germans. Had this happened, it is suggested, all that followed would have been different. This supposition is highly questionable. As long as Germany by 1900 was determined to build a fleet which, whatever its intent, plainly challenged Britain's command of the seas and so its basic ability to feed itself, and as long as Germany in 1914 made the strike into western Europe and towards the Channel ports which it actually carried out, nothing about Britain's response was likely to have been altered by prior diplomatic arrangements.

Sir Edward Grey, Britain's Foreign Minister from 1905 to 1916, continued the reorientation of British policy in an attempt to deter Germany from continental aggrandizement. His generally balanced conduct of affairs has since been represented to attribute to him a major responsibility for Europe's lapse into war.

Again, we may note the case of Italy. That country had entered into the Triple Alliance with Germany and Austria–Hungary in 1882 at a time when it possessed territorial ambitions at the expense of France. By the early twentieth century Italy's ambitions lay elsewhere. So when war broke out in 1914 Italy did not take the side of its allies. That was not because Italian statesmen acted on different principles from the rulers of other European countries. They acted on the same principles: that alliances controlled action only when they embodied a nation's vital interests. Otherwise they did not.

The field of possible explanations is perceptibly narrowing. We are being driven to take into account what actually happened in the events which precipitated the clash of arms, and to ask what they reveal about causation and responsibility. Certain long-term facts need to be laid bare. From 1870–71 France existed in a condition of resentment towards Germany on account of the territory which had been torn from it in the aftermath of the Franco-Prussian war. It also existed in a condition of deepening anxiety towards Germany, as the conclusiveness of its defeat in 1870 was reaffirmed by the widening gap between the two nations as evidenced in Germany's expanding population, mounting economic power and ongoing military capacity. In these circumstances, and as Germany's foreign policy in the post-Bismarckian era grew ever more wayward and unpredictable, France could only gain any sense of security vis-à-vis its neighbour both by marshalling its own, smaller military resources and by securing powerful friends. By 1914 it had certainly secured an ally in Russia and a sympathetic friend in Britain. But these hardly cancelled out France's inferiority or vulnerability to possible German aggression. For Britain's military capacity was trivial by continental standards, and Russia's was rudimentary. Neither country in their recent military forays (against the Boers and the Japanese respectively) had managed to put up a convincing display even against non-European powers.

Instability in the Balkans (1): the Balkan Wars of 1912 and 1913 threatened the stability of the Austro-Hungarian empire and aggravated rivalry between the smaller states of this region. Illustrated are Bulgarian artillery officers in the war of 1913.

The uncertainty of great power relationships in the aftermath of German unification and Bismarck's supersession was complicated by the unstable circumstances of southern and eastern Europe. Russia's war with Japan had highlighted the internal instability threatening the tsarist regime and had dealt a savage rebuff to Russian expansion towards the Pacific. And Austria–Hungary, especially in the aftermath of Turkey's exclusion from Europe following the Balkan war of 1912, found its whole multi-national structure challenged by the principle of national identity and by the emergence of Balkan states embodying that principle. Russia's partial identification with these states rendered that challenge all the more ominous.

The position of Britain was singular. Britain

had appeared during the nineteenth century largely absorbed with its expanding economy and empire, and happy to adopt a posture of splendid isolation towards the continent of Europe. Yet there was a large element of unreality here, which by 1900 Britain's rulers could no longer disregard. The fundamental concern of Britain's foreign policy remained what it had long been: the European power balance. The emergence of any over-mighty state, capable of dominating the land

mass of western and central Europe, of occupying the Channel ports (thereby disrupting Britain's commerce with Europe), and of challenging Britain's control of the sea approaches to its ports, constituted a direct threat to Britain's survival as an independent, self-respecting state. And during the nineteenth century, as Britain became more heavily populated, as its capacity to feed itself diminished dramatically, and as its economy grew ever more dependent on overseas commerce, these perils to its survival came to surpass any menace it had confronted in the times of Philip II, Louis XIV or Napoleon.

Instability in the Balkans (2): Archduke Franz Ferdinand, whose assassination by a Bosnian nationalist on 28 June 1914 set in train (though hardly caused) the sequence of events which culminated in the outbreak of a European war.

For much of the nineteenth century, no large power embodying these perils was in existence. Hence Britain could indulge in the luxury of splendid isolation. By 1900 this had changed. Germany's decision to enhance its powerful economy and its overweening military power by the creation of a great navy – a navy devoid of serious ocean-going capacity and so possessing no evident purpose except that of menacing Britain's ability to sustain itself – rendered splendid isolation no longer an option.

What these circumstances of the nations of Europe amounted to was a situation in which a great risk of conflict existed, but where no overwhelming compulsion towards the employment of unbridled violence was present. Indeed, many forces acted against any resort to war. The internal fragility of some governing regimes, the unhappy military experiences of others, the menace posed to all by the widespread accumulation of terrible weaponry, all pointed away from any lapse into savagery between nations. So also did the powerful urges towards international conciliation embodied in liberal ideals and socialist movements and the growing interdependence of industrial economies.

Yet, in the event, one great power did initiate and press forward upon a course of action the most likely consequence of which was a Europe-wide conflict. In the diplomatic flare-up of July and early August 1914, several powers – and most evidently Austria–Hungary and Russia – acted in ways that moved the crisis forward. But only one power first took action to initiate a crisis of international dimensions, then repudiated opportunities to bring the potential contestants to the conference table, and finally imposed on one major nation after another the harsh alternatives of either submitting tamely to drastic humiliation or of resorting to mobilization and then armed resistance.

In comprehending the eruption of July 1914, all roads lead to Berlin. Following the murder of the heir to the Habsburg throne on Austrian territory by a Bosnian nationalist, unwise heads in Vienna and Budapest may have contemplated savage action against Serbia, whose government was believed to be behind the plot. But the Austro-Hungarian authorities were bound to be kept in check by the likelihood of the tsar's response to the proposed elimination of a friendly Balkan state, unless Berlin provided unqualified assurance that Russia's response might be disregarded, because Germany's armed might would – if need be – deal with it. Only the action of the kaiser's government caused Vienna to draw up an ultimatum deliberately unacceptable to Belgrade, intended not as an act of diplomacy but as a prelude to invasion. And only pressure by the German government kept Vienna to this course when the Habsburg authorities showed signs of hesitation. The kaiser's revealing comment in the last days of the crisis, that Belgrade's near-acceptance of the Austro-Hungarian ultimatum removed all cause for war, did not affect Berlin's thrusting forward of Austria–Hungary into a prompt invasion of Serbia.

These events were bound to produce a response from Russia, the very least of which was likely to be the mobilization of some or all of its forces within its own

The sword and the sceptre. The German kaiser with General Paul von Hindenburg symbolizes the military emphasis of Germany's dominant forces and state structure.

borders. Yet when the rulers of Germany set that response in train, they knew full well what action on their part would follow. They would not just declare war upon Russia. Even more importantly, they would embark upon massive acts of armed aggression against both Belgium and France. (That such aggression would have consequences as regards British involvement in a European conflict could not be in doubt.)

That is, a German declaration of war upon Russia, for which Russian mobilization would provide a sort of justification, would instantly be followed by a devastating pre-emptive strike by German forces against France, delivered through the heartland of Belgium. This military action would instantly circumvent the line of fortresses carefully constructed by the French on their border with Germany and would cut Britain off from easy access to the continent. In six weeks, according to the designs of the German military, Belgium would be overrun, the Channel ports occupied, Paris taken, and the principal

OPPOSITE: *The rival alliances (2): the tsar receives a French military mission in August 1913.*

Tsar Nicholas II of Russia, the sadly under-equipped ruler whose unresponsiveness to progressive forces within his country contributed to internal tensions before the war and military calamity during it.

French armies ground to pieces on their frontier by German forces coming at them from two directions. Thereafter, the under-munitioned armies of Russia and any hostile Balkan states would be demolished at leisure. As for Britain, its considerable navy and insignificant army would have no meaningful place in these powerful events.

For the men who actually controlled power in Germany – not the Reichstag or the Social Democrat party, but Prussian aristocrats and the military élite, great industrialists and big-navy advocates – this plunge into war in both the east and the west, whatever its perils, contained huge attractions. The possibilities posed by German expansionism seemed endless: acquisitions of territory or the exercise of dominance in Belgium and northern France, in Russian Poland and the Ukraine, in the Balkans, and on through Turkey to Baghdad.

The great stumbling blocks to such acquisitions were liberal and socialist forces within Germany, the imperilled condition of the Austro-Hungarian empire, and the mounting coalition of powers prepared to resist Germany's threatened expansionism. Ruthless action by Berlin in July–August 1914 promised to eliminate all such obstacles in short order. Potential dissenters at home, particularly in the Social Democrat party, would be confronted with a war between German culture on the one hand and tsarist backwardness on the other. Russia's halting efforts to turn itself into a twentieth-century military power would be cut short before they had had time to develop. German intervention in the Balkans would eliminate the perils confronting the Habsburg empire. As for the potential menace of a Franco-Russian alliance constraining German ambitions by the prospect of a war on two fronts, that would never come to fruition: the Western Front would have been eliminated before the Eastern Front could lumber into action. And any possibility that the Anglo-French entente might become a military reality (a possibility up to now rendered theoretical by Britain's stubborn refusal to raise a mass army), that would be eliminated in a matter of weeks.

To reiterate: the outbreak of a great war may be attributed to a complex accumulation of events located well back in the historical past. But it may also be attributed to terrible simplicities: to a limited group of men in a particular country, in control of the levers of power and confident of the obedience of their subjects, acting to seize upon large and tempting objectives and none too inhibited by the reflection that, should calculations fail, frightful consequences might follow. So it was in 1914.

CHAPTER ONE

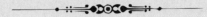

1914

GERMAN TROOPS ON THE EASTERN FRONT. *Although the war in the East now took on the static nature of the conflict on the Western Front, there were some chance elements in common, two of which are displayed here: barbed wire and machine-guns.*

1914

THE PLANS OF WAR

All the major continental powers had made large military plans for the eventuality of conflict, be it a war that they had provoked or one that had been forced upon them. Each plan was designed to secure speedy victory over at least one major adversary.

Germany intended to direct the bulk of its military effort westward. For its Eastern Front, it planned only a holding action against the Russians, on the assumption that the tsar's forces could be dealt with once the main opponent – France – had been knocked out. In the west, German forces would be employed in accordance with the inspiration of Count Alfred von Schlieffen, former chief of the German general staff (retired 1905), whose ambition was a decisive victory over France in the space of a mere six weeks. The weight of the German army's thrust would be placed on its right wing and directed initially against Belgium, whose rapid conquest would be preliminary to a great sweep south towards Paris. In this way the kaiser's forces would circumvent the whole defensive system on the Franco-German border constructed by the French in the aftermath of the catastrophe of 1870–71. And there would be convenient by-products. Belgium would henceforth be rendered a German satellite. And any response Britain might make to the destruction of a small country whose existence it regarded – with good reason – as vital to its security would be too insignificant to matter.

Alfred von Schlieffen, chief of the German general staff from 1891 to 1905 and author of the grandiose plan for swift military victory over France which was implemented in its essentials in 1914.

The German armies invading France would not, according to Schlieffen's scheme, halt when they had taken Paris. They would swing east and drive the French forces on the border into the waiting embrace of Germany's more southerly armies. Thereby the elimination of France and the kaiser's domination of western Europe would be total. Clearly, if all went according to this ambitious plan, the opening weeks of the war would – from the viewpoint of Berlin – have been well spent.

The military plans of France for the coming of war were hardly as ambitious as this, but they possessed a major common quality: the emphasis on offence. In the years immediately following 1871 and in the aftermath of military disaster, French army planners had thought defensively. They had constructed great fortresses on the border with Germany to fling back the invader. Only once that had been accomplished might they move to advance. But by 1914 the offensive had become the prevailing orthodoxy and élan the expected key to victory. Joseph Joffre, since 1911 the French commander-in-chief, intended (in accordance with his Plan 17) to propel his central armies into the lost provinces of

Alsace–Lorraine, drive the Germans back against the Rhine, and then swing north to cut the lines of communication of German forces coming through southern Belgium.

As the last aspect of Joffre's plan suggests, he was not totally disregarding the fact that the enemy would be making a major strike on his left. But he quite failed to appreciate how wide-ranging would be the Germans' intended sweep in the north, and concluded that they would confine their intrusion into Belgium to the one-third of that country south of the River Meuse. His miscalculation was compounded by another. Joffre failed to realize that the Germans would immediately introduce into the battle not just their active but their reserve divisions, and so be able to act with more power on more parts of the front than French forces were disposed to deal with.

Russia, as much as Germany, faced the dilemma of confronting two major opponents on two quite separate fronts. But unlike Germany, the tsar's advisers could not bring themselves to throw the main section of their forces against one front and opt for defence on the other. There was a case for striking hard against Germany, while its attention was turned west. Alternatively, there was a case for taking advantage of Germany's defensive posture in the east to thrust against Austria–Hungary, as constituting the more fragile opponent and as embracing the region where Russia's ambitions lay. What there seemed no case for doing was acting aggressively on both fronts. Russia lacked the weaponry and structure of command for war-making on such a scale. And as its railway system ran predominantly east–west, while the two fronts on which it proposed to act were respectively in the north and the south, it would be severely hampered in moving forces from one front to the other in accordance with military necessity or military opportunity. Nevertheless, the dilemma of choice proved too much for the tsarist High Command. It opted for two major offensives: into East Prussia in the north, and into Galicia (with a projected breakout into the plains of Hungary) in the south.

Austria–Hungary, equally, was intending simultaneous assaults on two foes, but in this case against unequal adversaries. Conrad von Hötzendorff, the Habsburg commander-in-chief, intended to deal once and for all with the menace

Joseph Joffre, French commander-in-chief, 1911–16. His obsession with the offensive brought him calamity in the opening weeks of campaigning (Plan 17), but he redeemed this by halting the German thrust on Paris at the battle of the Marne.

which the mere existence of Serbia was deemed to present to the survival of the Dual Monarchy. He would strike south across the Danube, occupy Belgrade and rout the Serb army. But his ambitions did not cease there. A second, much larger section of his forces would drive into southern Poland and settle with the forces of the tsar.

One great power, though soon involved in the conflict, had no large military plans. For one thing, it had no large army. Britain's peacetime force was a small, professional army, consisting only of volunteers, and had been designed to deal with disturbances outside Europe. (Sometimes, as the Boer War had demonstrated, it could manage even that only with difficulty.) This meant that in 1914 Britain simply had no capacity to engage in an independent campaign against any of the great conscript armies of continental Europe. So as early in the twentieth century, Britain grew increasingly aware of the menace posed by Germany's combination of military, industrial and naval might, it responded by becoming more closely associated (but not allied) with France. One facet of this was the decision that, in the event of Germany's instituting a general western European conflict, the small British Expeditionary Force (BEF) would most likely act as an adjunct of the army of France. That is, the BEF (consisting of one cavalry and six infantry divisions) would form up on the extreme left of the French line. According to French military calculations, even if part of Belgium was invaded, not much fighting would occur here.

In so far as Britain was expected to make a major contribution to the early months of a European war, it would not be on land. The mere existence of a mighty German battlefleet presented Britain with a challenge which, if allowed to come to fruition, would promptly render it powerless to survive as a self-sustaining state. Britain's major response to this deadly threat must be the assemblage in the North Sea of its principal naval resources. Thereby it would force the German fleet to remain inactively in harbour, or it would bring it to battle on the high seas in circumstances where, after years of investment and training, the British possessed a clear predominance in numbers and weaponry.

Conrad von Hötzendorff, chief of the Austro-Hungarian general staff 1906–17, whose failure to overrun Serbia in the opening encounters of the war proved a prelude to a succession of defeats on the Eastern Front, but a stubborn and generally successful defence against Italy.

The plans in action

How did these plans work out? Which prospered, and which went awry? Were any of them, for that matter, soundly based?

Austria–Hungary, having pushed Europe – if only in response to reckless encouragement from Berlin – over the brink, started the business of actually waging war. As a beginning, Habsburg forces crossed the Danube and occupied Belgrade. This proved a confused undertaking. A section of the units which started south was then, without engaging the enemy, redirected to the front against Russia, with numerous complications in the matter of transportation. As for that section of the Austro-Hungarian army which continued the invasion, it

Lacking a large conscript army in 1914, Britain swiftly began the transition to a major military power by enrolling a great volunteer army. The variety of headgear worn by the potential soldiers indicates the width of the social spectrum from which they were drawn.

THE SCHLIEFFEN PLAN

The rival plans and their outcomes. The projected sweep by the Germans through Belgium and down to the west of Paris was amended during its progress.

The Schlieffen Plan
1914

- planned German attacks
- German army positions
- original Plan 17 20 May 1913

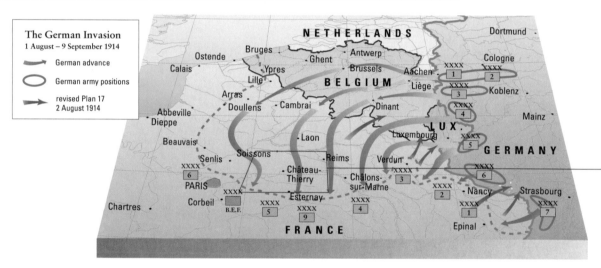

The German Invasion
1 August – 9 September 1914

- German advance
- German army positions
- revised Plan 17 2 August 1914

speedily came to grief. A devoted Serbian counter-attack liberated Belgrade and threw the invaders back across the Danube. (In November–December 1914 Conrad's forces repeated the assault on Serbia, with identical results.) For the Habsburg empire it was not a good beginning.

On the Carpathian front, engaging the Russians, Conrad's experiences were no happier. Again, there was an initial impressive advance. Again, there was counter-attack and retreat. The advance sprang from over-confidence. Conrad sent his forces forward without taking the precaution of ensuring adequate fire support and without requiring his armies to maintain a continuous front. Outnumbered by the Russians (and no better equipped), the Habsburg forces soon found themselves suffering heavy casualties and in danger of being surrounded. This led to a rapid retreat.

So in the first half of September 1914 the Austro-Hungarians not only

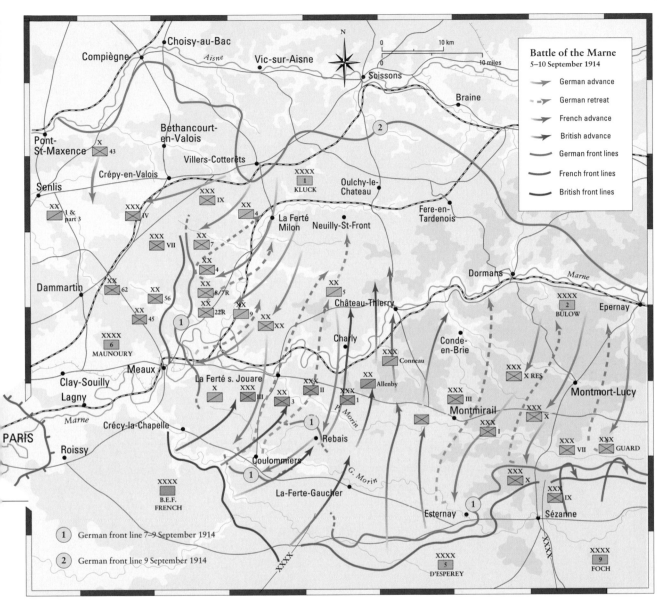

1 German front line 7–9 September 1914
2 German front line 9 September 1914

abandoned territory they had just overrun but suffered heavy losses in manpower. Some one-third of a million of their men became casualties and another 100,000 were taken prisoner. Particularly heavy were the losses among junior officers and NCOs. This was of great moment. The morale of what was in effect a multi-national army, whose allegiance was not to a nation or a unitary state but to a common sovereign, depended heavily on the qualities of its lower levels of command. The pre-war cadre of junior officers and NCOs had familiarized themselves with a variety of languages, and had seen it as their task to maintain rapport with a diverse body of rank and file. When they were gone there were no adequate substitutes. What came in their stead were middle-class Austro-Hungarians, displaying the attitudes usual among dominant nationalities in command of people they perceived as inferior stock.

As is apparent from the above, Russia's forces scored imposing successes in

BATTLE OF THE MARNE

The map depicts the turning point of the battle, as the French and British took advantage of the developing gap between Kluck's First and Bülow's Second Armies to drive between them and force their retreat.

August and September 1914 on their southern front. In the north, attacking into East Prussia, it was a different story. Here the superior qualities of the German state structure, industrial base, weaponry and Command personnel told decisively against the tsar's armies. This would not be a transitory phenomenon. Despite subsequent heroic efforts among the progressive sectors of the Russian home front, the encounters between Germans and Russians in 1914 would prove to have set a pattern which would be repeated until the collapse of tsarism.

In the latter half of August 1914, the Russian First and Second Armies, superior in numbers – if not in equipment or military intelligence – to their adversaries, advanced into East Prussia. The two armies moved independently, one to the south of the Masurian Lakes and the other to the north. The ill-effects of this dispersal were compounded by the action of the Russian authorities in telegraphing their movements in unencoded messages. Given the vast distances that characterized the Eastern Front, this failure to employ codes was not peculiar to the Russians. But only their adversaries possessed the organization to take advantage of the information received.

As a consequence, the German eastern command, which was in the process of coming under the direction of the soon-to-be-famous team of Hindenburg and Ludendorff, discovered the whereabouts of the two Russian armies. The Germans acted promptly to offset their own numerical disadvantage by engaging the opposing forces separately. At Tannenberg in East Prussia, in an action opening on 26 August, the Russian Second Army found itself surrounded and cut off from supply and reinforcement. It surrendered. For almost trivial losses by the Germans, a Russian force of 125,000 had simply vanished from the scene.

Its supposed partner, the Russian First Army, did not fare quite this badly. Its commander maintained a front to the enemy and avoided encirclement. But in the face of the decided superiority now possessed by the Germans opposing him, he had no choice but to fall back

FRENCH M1897 75MM FIELD GUN

The principle of axial recoil is demonstrated by the action of the barrel of the gun when firing. This obviated the need to reposition the gun between rounds and so greatly increased the rate of fire. The recoil action of *the French M1897 75mm field gun was the first design to control this aspect of modern artillery successfully, and was thus the prototype of all field artillery pieces used in the First World War.*

from East Prussia. So ended – although none could be certain of this at the time – any occupation of German territory by hostile forces for the remainder of the war.

While Russian schemes for a great sweep into Germany were ending in calamity, the French on the far side of Europe were faring no better. Joffre's much-vaunted Plan 17 for the liberation of Alsace–Lorraine and advance to the Rhine swiftly came to grief. German preparations to resist invasion had been well made. French plans to carry it out had not. Ineffectively commanded, badly supplied with such necessities as maps of the areas to be overrun, and often exposed to the enemy by their brightly coloured uniforms, Joffre's armies did not engage in the headlong dash to glory that has sometimes been portrayed. Their operations were rather a poorly co-ordinated stumbling forward into ill-charted territory and against well-sited defences. In the space of just over two weeks, the French attackers suffered a third of a million casualties and found themselves back in their original positions, except for a toehold in Alsace.

Germany's new command in the east. Erich Ludendorff, following his success against the Belgian fortress of Liège, was put in effective command against the Russians, with the hitherto retired Paul von Hindenburg as his nominal chief.

Street fighting in Hohenstein in August 1914 during the Russian invasion of East Prussia.

Battle of Tannenberg, 26–31 August 1914. A view overlooking the ruins of the town after the battle which saw the destruction of the Russian Second Army.

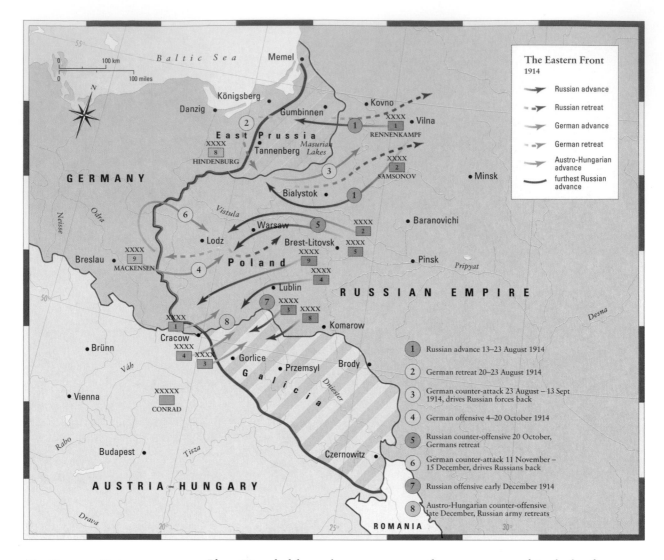

The Eastern Front
1914

	Russian advance
	Russian retreat
	German advance
	German retreat
	Austro-Hungarian advance
	furthest Russian advance

1 Russian advance 13–23 August 1914

2 German retreat 20–23 August 1914

3 German counter-attack 23 August – 13 Sept 1914, drives Russian forces back

4 German offensive 4–20 October 1914

5 Russian counter-offensive 20 October, Germans retreat

6 German counter-attack 11 November – 15 December, drives Russians back

7 Russian offensive early December 1914

8 Austro-Hungarian counter-offensive late December, Russian army retreats

THE EASTERN FRONT,
AUGUST–DECEMBER 1914

*This map illustrates the
fluctuating fortunes of
Russia's forces in the twin
campaigns against the
Central Powers.*

Plan 17 ended here, but not necessarily on account of its lack of success. Joffre was responding to ominous events on the far left of his front. In sharp contrast to the French débâcle, Germany's Schlieffen Plan was proceeding with terrifying success. In the opening days of the conflict, the kaiser's divisions smashed through the great Belgian fortresses and swept aside the resistance of the Belgian army. Attempts by British and French troops to halt the onrush succumbed to sheer weight of numbers. By 23 August the Allied left wing was in full (if orderly) retreat. For thirteen days the German First and Second Armies on the right of Schlieffen's great wheel drove southward into France until they were across the River Marne and within sight of Paris. Apparently, 1914 would prove a re-run of 1870.

But appearances were deceptive. In 1870, by the time the Prussian armies reached Paris the main forces of France had already been overwhelmed on the battlefield. In August 1914 this was yet to be accomplished. Joffre's armies, where they had taken the initiative on the central and southern sectors, had certainly sustained a heavy rebuff, but they were maintaining an unbroken line. (The

The great retreat of the
French and British forces
from Belgium and northern
France from 23 August left
the obsolete fortress of
Maubeuge isolated. It fell
to the Germans on
7 September 1914.

Helmuth von Moltke the
Younger, German
commander-in-chief on the
outbreak of war,
endeavoured to put into
practice a modified version
of the Schlieffen Plan. The
failure of this enterprise,
signalled by the Franco-
British victory on the Marne
on 6 September 1914, led to
his breakdown and
replacement.

Germans in these sectors were also moving on to the offensive.) And in the north, which to the Germans was the critical front, the French, while falling back, had suffered no compelling battlefield defeat. The Allied flank had not been turned, nor the armies of France overwhelmed.

The scene was thereby set for a striking change of fortune. This is not to be accounted for, as is often claimed, by the fact that Helmuth von Moltke the Younger (the German c-in-c who had succeeded Schlieffen) fatally amended the entirely viable plan provided by his predecessor. The plan was less than viable. Schlieffen, in truth, had never done the spadework regarding logistics and transportation which alone might reveal whether so bold a scheme could ever be made to work. In the event, only because the Germans chose to attack at harvest time were their soldiers even able to feed themselves on their journey south. (Thousands of their invaluable packhorses, on which supply of weapons depended, were less fortunate. They died of starvation and exhaustion.) And the distances that the kaiser's rightward armies had to travel reduced the soldiery to a state of collapse and stretched their supply lines to breaking point. So even had the German First and Second Armies managed to encircle Paris and fight a successful action at the Marne, it is unlikely that they could have forthwith set off east to the French frontier and disposed of Joffre's forces there.

As it happened, a significant proportion of the French forces which had been engaged in Joffre's initial attacks were no longer to be found on the frontier. Among the shortcomings of Schlieffen's plan was its assumption that the armies of its opponents would remain in the positions required of them to ensure the plan's success. Such assumptions are not always unwarranted. France had obliged its enemy well enough in 1870, and

The coming of stalemate. Driven into retreat, the Germans only fell back to high ground above the River Aisne, where they dug trenches.

would do so again in 1940; 1914 proved to be different. Whatever the early blunders of Joffre, he responded to the German advance on Paris coolly and appropriately. He called off his own offensives, placed on the defensive the forces he calculated it was necessary to leave on the Franco-German frontier, and set about moving a significant proportion of his troops from the frontier to the French capital.

This transformed the strategic situation. As the German right wing, still fully engaged by the French and British forces it had driven into retreat, crossed the Marne and brushed past Paris to its west, it found itself quite outmanoeuvred. Joffre's newly formed Sixth Army struck at the enemy's flank, forcing the German First Army to swing towards Paris. At this moment the hitherto retreating French forces counter-attacked Moltke's Second Army. An ominous gap opened between the two German forces. Into it the French, supported by the BEF, began to thrust. To save their right wing from calamity, the German Command had no choice but to pull back and re-form. As they did so, the castles which Schlieffen had sculpted so carefully in the air crumbled to nothing.

THE ONSET OF STALEMATE

The retreat of the kaiser's forces from the Marne was decisive in one sense. Germany's plans for a thunderclap victory in the west were dead. In all other respects it was anything but decisive. Far from making a headlong flight to German territory, as some in the Allied Command anticipated, the German right wing fell back only to the River Aisne and the high ground above it. There it halted, dug trenches, sited machine-guns, set up barbed-wire entanglements and placed artillery in position. This proceeding, although the fact could hardly be appreciated at the time, was a terrible foreshadowing of Western Front conflicts for the ensuing three years. Germany's armies, when acting on the defensive in the west – as, with one important exception, would be the case until the end of 1917 – withdrew to accommodating high ground, developed an extensive trench system, and assembled large quantities of defensive weaponry, especially artillery.

The Aisne in mid September 1914 proclaimed the shape of things to come. French and British forces struggled in vain to take the high ground above the river. On 17 September, four days after the engagement began, a senior British officer recorded: 'We were fairly quiet but we are in a stalemate.' This remark told more about the ensuing form of the war in the west than its author could possibly have recognized.

By late September, stalemate had become the situation along the whole Western Front from the River Aisne south to the Swiss border. In that long stretch of territory the armies of France and Britain on the one hand, and of Germany on the other, had reached deadlock. If a war of movement was still on offer, it could only occur north of the Aisne, by one side swinging beyond the flank of the other and so threatening to sever its opponent's lines of communication.

At first rudimentary, as depicted by the British (below), trenches became increasingly formidable as battle proceeded, as evidenced by the German trench (opposite), its occupants still wearing soft headgear.

So there developed a succession of sharp moves as each side endeavoured to outflank its opponents. These actions are known, on account of their ever more northerly flow, as the 'Race to the Sea'. Neither side, in truth, was seeking to get to the sea. But as each attempt at a swing north was countered by a comparable move on the other side, the coast at last came in sight and no flank remained available for the turning. These successive movements were accompanied, on the part of whichever adversary was thrust on the defensive, by a repetition of the actions of the Germans on the Aisne: the speedy construction of an effective

In one of the first German offensives on the Eastern Front, members of Hindenburg's army approach the rail centre of Lodz on their way to Warsaw (then part of Russian Poland).

defensive system consisting of trenches, barbed wire, machine-guns, artillery and sufficient riflemen to hold off an attack. The expression 'we are in a stalemate' was coming to apply not just to one sector of the trench line which the ebb and flow of battle had thrown up, but to a continuous barrier of earthworks, weaponry and infantrymen stretching all the way from the Belgian coast to the border of Switzerland.

THE PERSISTENCE OF STALEMATE

The demise of the Schlieffen Plan had shattered the nerve of Moltke and caused his replacement by Erich von Falkenhayn, former Prussian Minister of War.

The situation confronting Falkenhayn between October 1914 and the end of the year was profoundly ambivalent. Germany had scored major victories over the Russians and driven them from Prussian territory. But although the commanders in the east, Hindenburg and Ludendorff, saw this as an opening for a great campaign in their region, Falkenhayn was not attracted. Certainly, with Austro-Hungarian forces in steady retreat before Russia's successful southern offensive, he had no choice but to divert German divisions to the south-east, thereby obliging the Russian command to move the bulk of its forces northward, away from the Austrians, to confront the Germans in the vicinity of Warsaw. So the pressure on Germany's struggling ally was relieved. But more than this Falkenhayn was not eager to do. A campaign in the east, because of the vast spaces, seemed to lack an attainable objective. And it would not be directed against Germany's principal adversaries.

Victory, for Falkenhayn, was synonymous with triumph in the west: the elimination of the present military power of France, and the negation of the emerging military power of Britain. The failure of the Schlieffen Plan had removed the prospect of total victory over France before the end of the year. And there was no likelihood of crippling Britain in short order, given that the kaiser would not contemplate sending his battlefleet into the open sea to risk a full-scale engagement with the British navy. So France would remain a viable military power when 1915 dawned. And the British fleet, although denied the climactic sea battle it craved, would continue to command the North Sea and English Channel, would maintain Britain's vital access to the raw materials and commerce of the outside world, and would be free to mount an increasingly effective sea blockade of Germany.

Nevertheless, Falkenhayn concluded that an operation seriously damaging to the military prospects of France and Britain could be launched in what remained of 1914. His purpose was to force his way through to the ports of Calais and Dunkirk, thereby winning a significant battlefield victory and denying Britain its most direct route to France and Belgium. So in mid October, in that small part of Belgium that remained in Allied hands, and employing both veterans and eager young recruits supported by the massive artillery which had destroyed the Belgian forts, he launched a savage offensive. It has gone down in history as the First Battle of Ypres.

The assault fell mainly on the British Expeditionary Force ('the Old Contemptibles', after a reference by the kaiser to Britain's 'contemptible little army'). Heavily outnumbered in men, hugely outnumbered in weaponry, and

Erich von Falkenhayn, Moltke's successor as German commander-in-chief. Convinced that victory lay in the west, he launched the bloody offensive towards the Channel ports which was brought to a halt in the First Battle of Ypres (30 October – 24 November 1914).

The prelude to the First Battle of Ypres. British troops, intent on turning the German flank, advance across a Belgian field on 13 October 1914. Within days they would be assailed by Falkenhayn's forces.

occupying only a succession of shallow, usually waterlogged trenches, the BEF suffered heavy casualties and was gradually driven back towards the town of Ypres. Yet Falkenhayn failed utterly in his purpose. The losses among his youthful recruits proved too devastating to be borne. And the Channel ports remained beyond his grasp.

Joffre also was looking for a sort of success with which to end the year. While Falkenhayn attacked in the north, Joffre, encouraged by the removal of German divisions to the Russian front, launched offensives in the Champagne and the Vosges – the centre and extreme south of the Western Front. His purpose was to drive German forces back across their own border.

As late as 25 December (for although informal Christmas truces sprang up along many parts of the front, these were not universal) Joffre sent his forces forward against strongly entrenched opponents in conditions of blinding snow, low cloud, fog and mud. Their endeavours availed nothing. Like First Ypres, they simply reinforced the grim message that had proclaimed itself in the west since the Aisne in September: that trench defences serviced by ample manpower and sufficient weaponry possessed a terrible capacity to repel attack.

Germans firing an Austrian-made heavy howitzer near Ypres, 1914. Despite such heavy weapons, the Germans were unable to dislodge the British from their hold on Ypres.

Le Petit Journal

ADMINISTRATION
61, RUE LAFAYETTE, 61

Les manuscrits ne sont pas rendus

On s'abonne sans frais
dans tous les bureaux de poste

5 CENT.   SUPPLÉMENT ILLUSTRÉ   **5** CENT.

27me Année ——— ✦✦ ——— Numéro 1.307

DIMANCHE 9 JANVIER 1916

ABONNEMENTS

	SIX MOIS	UN AN
SEINE et SEINE-ET-OISE..	2 fr.	3 fr. 50
DÉPARTEMENTS..........	2 fr.	4 fr. »
ÉTRANGER	2 50	5 fr. »

LE GÉNÉRAL HIVER

CHAPTER TWO

1915

GENERAL WINTER. The campaigning season remained a vital aspect of the First World War. Despite the reluctance of some commanders to acknowledge the fact, military operations of any substance proved impossible while 'General Winter' commanded the field. As depicted in this French publication, the monstrous apparition of winter renders infantry, cavalry and artillery equally useless.

1915

THE LIMITS ON CHOICE

As 1915 dawned, the substantial failure of the opening moves appeared to point in one of two directions. Either the war should be called off. Or the combatants should abandon their costly attacks and resort to the defensive.

On all sides, such conclusions were deemed unacceptable. The great autocracies of central and eastern Europe would be seriously challenged from within if they failed to press on to the accomplishment of a compelling victory. In too many ways, the validity of these regimes depended on success in war. As for the liberal powers of the West, they had entered the struggle for clearly defined purposes which would be negated if they either abandoned the conflict or fell back on the defensive. Those purposes were the restoration of the independence of Belgium, then almost totally under German subjection, and the assertion of the territorial integrity and security of France, one-tenth of whose soil (including a third of its industrial region) was in enemy hands.

Yet if objective circumstances required every participant to resume the attack in 1915, large questions presented themselves. Where should these operations take place? And how were they to be rendered more effective than the failed endeavours of 1914?

FALKENHAYN'S DILEMMA

The choices facing Germany were particularly taxing. Falkenhayn, as before, wished to strike in the west, where alone lay ultimate victory. But pressure to act on his Eastern Front was strong. Hindenburg and Ludendorff were determined

'Battles against the Russians in the Carpathians'. A picture postcard (photomontage against painted background) reflecting the bizarre representations of war in popular imagery.

on a further campaign in Masuria, and did not hesitate to draw the kaiser's attention to the contrast between their own successes in that region in 1914 and Falkenhayn's costly failure at Ypres. Further south, the Austro-Hungarian commander-in-chief was making plain the straitened state of his forces, and hinting to his ally that without substantial reinforcement his line in the Carpathians might not hold.

Falkenhayn, anyway for the moment, could not disregard the combination of promise and threat presented by the Eastern Front. With evident reluctance he decided to stand on the defensive in the west and make his forward moves against the Russians. Had he but known it, he was committing himself to two eastern offensives – one in Masuria, which was the choice of Hindenburg and Ludendorff, and one in the Carpathians, to which Conrad was determined to commit him. The Austro-Hungarian commander, in truth, was more sanguine about the state of his army than he was admitting to Falkenhayn. But he was also mindful of the need to relieve the fortress of Przemysl, where, surrounded by the Russians, 120,000 of his troops were holding out. Conrad intended to employ German reinforcements, not to shore up his line, but to complement his own forces in a drive out of the Carpathians towards Przemysl.

The Russians also had to make a decision concerning strategy. Should they launch a further offensive in East Prussia? Or should they attack Conrad's forces and attempt a breakout into the Hungarian plain? In the event they hesitated too long. The Central Powers struck before any of their preparations were completed.

The great Austro-Hungarian fortress of Przemysl, having fallen to the Russians in March 1915, is here seen being reoccupied as a result of the offensive launched from Gorlice–Tarnow in May 1915.

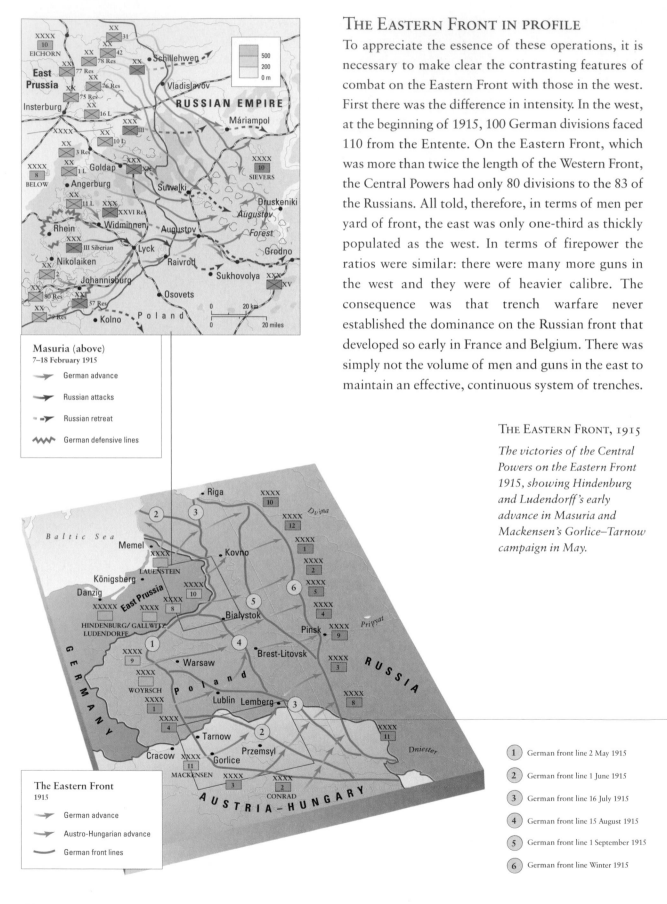

Masuria (above)
7–18 February 1915

→ German advance

→ Russian attacks

⇢ Russian retreat

〜〜〜 German defensive lines

THE EASTERN FRONT IN PROFILE

To appreciate the essence of these operations, it is necessary to make clear the contrasting features of combat on the Eastern Front with those in the west. First there was the difference in intensity. In the west, at the beginning of 1915, 100 German divisions faced 110 from the Entente. On the Eastern Front, which was more than twice the length of the Western Front, the Central Powers had only 80 divisions to the 83 of the Russians. All told, therefore, in terms of men per yard of front, the east was only one-third as thickly populated as the west. In terms of firepower the ratios were similar: there were many more guns in the west and they were of heavier calibre. The consequence was that trench warfare never established the dominance on the Russian front that developed so early in France and Belgium. There was simply not the volume of men and guns in the east to maintain an effective, continuous system of trenches.

THE EASTERN FRONT, 1915

The victories of the Central Powers on the Eastern Front 1915, showing Hindenburg and Ludendorff's early advance in Masuria and Mackensen's Gorlice–Tarnow campaign in May.

The Eastern Front
1915

→ German advance

→ Austro-Hungarian advance

— German front lines

1 German front line 2 May 1915

2 German front line 1 June 1915

3 German front line 16 July 1915

4 German front line 15 August 1915

5 German front line 1 September 1915

6 German front line Winter 1915

A further difference between the two fronts was to be found in the quality of communications. Roads and railways were markedly inferior in the east. On the Russian side it was particularly difficult to bring troops speedily from the rear. And the Russians had no lateral railways that could transport divisions quickly between their northern and southern fronts. The Germans were better served with road and rail communications behind their lines, but once they began advancing into enemy territory they encountered the Russian communications vacuum. Nor could they easily compensate for this by utilizing their own rolling stock. The rail beds were lacking, and German and Russian lines were of different gauge.

What therefore usually happened was that the Germans, aided by their superior artillery resources and with reserves placed close up to their front, managed to make substantial initial advances. Then they lost momentum as their supply problems became insurmountable and as the lumbering Russian machine at last fed in its reserves.

The consequence was curious. The Eastern Front saw many more battles of manoeuvre than the Western. But because of the vast distances and poor communications involved, these wide-ranging contests might prove as unproductive of any final decision as the small gains in territory

GORLICE–TARNOW

The Austro-German campaign launched between Gorlice and Tarnow in May–June 1915 brought to the Central Powers the most spectacular advances of the war. Nevertheless, these great victories proved insufficient to defeat the tsar's armies.

Gorlice–Tarnow
May–June 1915

→ Austro-Hungarian advance

→ German advance

〜 Austro-Hungarian front lines

〜 German front lines

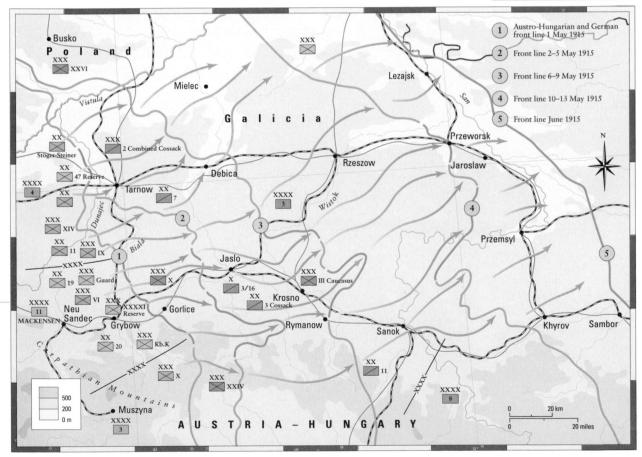

Russian troops near Minsk. Despite the heavy Russian losses in 1915 the tsar's army could still call on the services of a great pool of reserve forces.

which characterized the almost static battles in France and Belgium.

If, nevertheless, the Germans did get the better of the Russians to an extent no other combatant could manage on the Western Front up to the end of 1917, this was because of another great difference between the powers fighting in the east and in the west. In the west, the rival armies were roughly comparable in terms of organization and equipment. In the east, especially early on, the Germans were clearly superior to their adversaries in terms of artillery and machine-guns. And from start to finish they were superior in staff work – for example in their handling of reserves.

No such contrast, it needs to be added, applied to the Russian and Austro-Hungarian armies. Neither was amply equipped or blessed with a competent level of command. But the Russian army was a more cohesive force, unaffected by the nationality differences which increasingly afflicted the Habsburg army as the struggle proceeded. And in terms of raw manpower, Russia had a larger pool on which to draw as the established armies of both sides were swiftly eliminated. In these circumstances, the Russians were likely to outfight the Austro-Hungarians as long as they were free from outside interference. But this was never a situation that would obtain for long. Whenever the Habsburg forces became too hard pressed, those of the kaiser were bound to intervene.

GERMANY STRIKES EAST

The campaign in the Carpathians opened on 23 January 1915. Conrad initially managed some progress towards Przemysl. Then the terrain, rugged and under several feet of snow, prevented supplies from reaching his forces. In temperatures as low as minus 15 degrees centigrade, healthy soldiers froze to death and many of the wounded fell victim to wolves. By the end of the month Conrad's great design had come to a halt. Three other Carpathian offensives followed between January and March, two by the Russians and one by the Austro-Hungarians. All failed in their objectives – the Hungarian plain did not witness the arrival of the Cossacks, but neither was Przemysl relieved. Conrad's losses were prodigious. Most of his experienced officers and NCOs were now expended. Casualties totalled 400,000, with a further 120,000 (including nine generals) becoming prisoners of war when Przemysl fell on 23 March. From this period – although it seemed to escape Conrad's attention – the Austro-Hungarian army was without independent offensive capability. It could now only mount attacks in conjunction with the Germans. The Russians, despite their 400,000 casualties, had not yet quite reached this condition.

General August von Mackensen, commander of the Eleventh German Army in Galicia 1915. He was responsible for the plan which resulted in the spectacular successes of the Central Powers in that year.

As exhaustion settled over the Carpathians, Hindenburg and Ludendorff were attacking in the north. Surprise was complete. Although hampered by atrocious weather, on the left the Germans drove back the Russians some 70 miles. On the right they did less well, encountering a fresh force (the Russian Twelfth Army) being assembled by the Russian General Staff (Stavka) for its own projected attack. Stubborn resistance by these troops thwarted any chance of a major encirclement. Consequently, the Germans fell short of strategic victory. They had advanced from one side of the Masurian Lakes to the other, but that placed their supply lines through an extended swamp. Even so, they had enjoyed a considerable success. In addition to overrunning territory, the Germans sustained many fewer casualties than the 200,000 they inflicted on the Russians.

Despite these satisfactory events in the north, at the beginning of April Falkenhayn was still not free to turn his attention to the Western Front. The situation of Austria–Hungary was too precarious. Habsburg casualties since July 1914 now totalled two million. Italy was known to be negotiating with the Entente to enter the war against the Dual Monarchy. Conrad hinted darkly at a negotiated peace. To prop up his ally, Falkenhayn was forced to rush eight divisions east from the Western Front to form the nucleus of a new Eleventh Army under August von Mackensen. This force was concentrated in great secrecy to the north of the Carpathians between the towns of Gorlice and Tarnow.

On the front of attack the Central Powers mustered eighteen divisions: ten German to strike the main blow and eight Austro-Hungarian to provide flank protection. Opposing them were five and a half Russian divisions of poor quality. Falkenhayn aspired to shatter the Russian front, then turn north and drive the tsarist armies occupying the Polish salient against Hindenburg's forces in East Prussia. The date set for the attack was 2 May.

The offensive opened with a four-hour bombardment that flattened the Russian defences (in fact little more than shallow ditches with occasional strands of barbed wire). By the end of the day a 5-mile gap had been opened. By mid May Mackensen's troops were on the River San, an advance of 100 miles. On 3 June Przemysl was retaken. On 20 June the Russians ordered a general retirement from Galicia. By this time the Italians had declared war on Austria–Hungary and some of Conrad's forces were being diverted to the new front, but this brought no relief to the Russians. At the end of June Mackensen was across the Dniester. This was the signal for widening the offensive. On 12 July Hindenburg attacked in the north, and an advance opened on Warsaw. The city fell on 4–5 August. Brest-Litovsk followed on the 25th.

Ludendorff, for one, anticipated the end of the tsar's armies. But at this point the intractable nature of the Eastern Front reasserted itself. The Germans had advanced 300 miles. Their railheads,

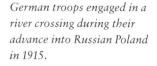

German troops engaged in a river crossing during their advance into Russian Poland in 1915.

pushed forward by the exertions of the railway troops, nevertheless lagged 50 to 100 miles behind the front. Supplies had to be manhandled the remaining distance. The troops too were becoming exhausted, the artillery was wearing out, ammunition was running short and water was proving hard to come by.

On the other side of the front the situation was mending. As the Russians retreated from the Polish salient, their line became shorter and fresh troops were more easily fed into the battle. Gradually the front stabilized. A new offensive into Courland in September, launched on his own initiative by Ludendorff, proved only a minor incident. Courland and Vilna were taken, but Riga – the real prize in this area – remained beyond Ludendorff's grasp.

Russian position in Grodno, August 1915. Soldiers with a heavy machine-gun in a shallow trench.

Overall, the Germans at Gorlice–Tarnow had achieved their greatest numerical success of the war. The Russians lost 850,000 in prisoners alone (140,000 by the third day of the campaign) and nearly a million were killed and wounded. By September some 3,000 Russian guns had been captured. German losses, although impossible to assess with any precision, were certainly far fewer. If total victory over the Russians was simply not attainable, any immediate danger from Russia was now past. Falkenhayn and Conrad were now free to turn their attention in other directions.

War in Italy: bringing up Austrian reinforcements. This illustration depicts the primitive state of warfare on the Italian front, brought about by the weather, the inhospitable terrain and the relative lack of equipment of the combatants.

THE ITALIAN CAMPAIGN

The campaign between Austria–Hungary and Italy. The unrewarding nature of the campaigns on this front is revealed in these two maps, one dealing with Austria's offensive on the Trentino in May–June 1916, the other with Italy's eleven offensives along the Isonzo between June 1915 and September 1917. The failure of the Italian army was due to Austrian possession of the commanding mountain heights.

A. The Trentino Offensive
May–June 1916

→ Austro-Hungarian attacks

⌒ Austro-Hungarian front lines

① Austro-Hungarian front line, 15 May 1916

② Austro-Hungarian front line, late June 1916

① Italian front line 15 June 1915

② Italian front line September 1917

B. The Isonzo Battles
June 1915 – September 1917

→ Italian attacks

⌒ Italian front lines

ITALY TO WAR

Meanwhile, a new front had opened. Italy had declared war against Austria–Hungary on 23 May. In one sense this was odd. Back in August 1914 Italy had been formally an ally of the Central Powers. But given its vulnerability to sea blockade and its dependence on Britain for its supply of coal, and with its territorial ambitions clearly directed against Austria–Hungary, there was never a chance of Italy entering the war against the Entente. Since then the inception of the Dardanelles operation seemed to herald a carve-up of the Turkish empire,

ITALY: DEFEAT AND VICTORY

The great Italian setback at Caporetto in October–November 1917, and Italy's success at Vittorio–Veneto in the last days of the war.

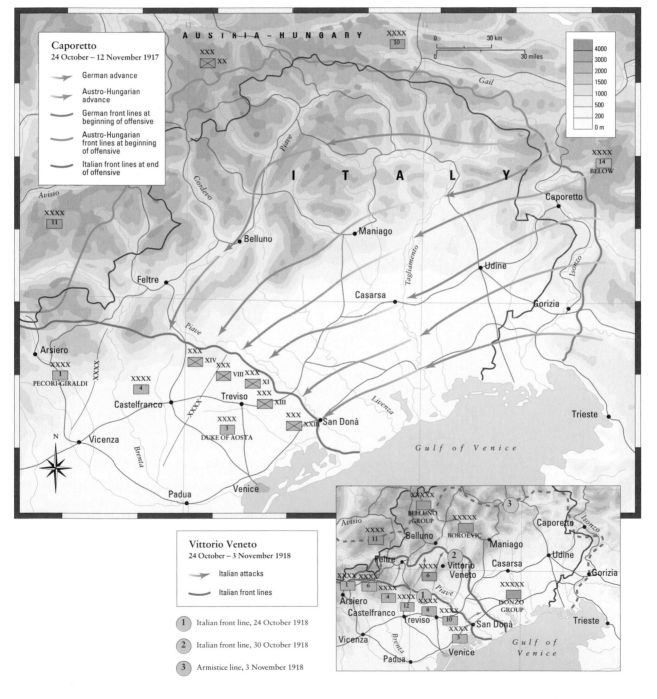

Caporetto
24 October – 12 November 1917

→ German advance

→ Austro-Hungarian advance

German front lines at beginning of offensive

Austro-Hungarian front lines at beginning of offensive

Italian front lines at end of offensive

Vittorio Veneto
24 October – 3 November 1918

→ Italian attacks

Italian front lines

(1) Italian front line, 24 October 1918

(2) Italian front line, 30 October 1918

(3) Armistice line, 3 November 1918

The unrelenting battles of the Isonzo. This photograph shows supplies being brought forward in spite of the unwelcoming terrain.

promising to the Italians – who had forcibly annexed the Turkish province of Libya in 1912 – further gains at Ottoman expense. Even the ominous successes of German forces against Russia could not act as a brake on Italian ambitions.

Yet Italy was ill-equipped to fight the war it chose to enter. The Italian army had not recovered from its exertions in Africa. It lacked modern artillery and was deficient in machine-guns. Moreover, of the 400-mile border which Italy shared with Austria, all but 20 miles were mountainous. The Austrians had taken up defensive positions not along the frontier but along the high ground within their own territory. Even the most promising area of attack, the plateau carved out by the River Isonzo from the sea to the lower reaches of the Alps, presented many obstacles. The Austrians occupied the high ground here also, and the slopes of the plateau were bare and rock-strewn.

Undeterred, the Italian government chose both to enter the war and to proceed promptly to the attack. General Luigi Cadorna (the Italian c-in-c) opted to launch his major campaigns of 1915 (in June, July and October) on the Isonzo. On each occasion the Italians outnumbered their adversaries in infantry and artillery, but these advantages were more than cancelled out by the nature of the terrain and by the fact that Italian artillery did not include a sufficiency of heavy guns. So by the end of the third battle the Italians had lost 125,000 casualties (as against 100,000 Austrians). For this heavy cost the Italian line had barely moved.

THE ELIMINATION OF SERBIA

Despite the immediate lack of Italian success, Austria–Hungary's position remained imperilled. And while the Gallipoli operations continued, Turkey also was under threat. From the German viewpoint, the Central Powers' position in the Balkans required attention. Serbia was the obvious target. Habsburg action against it in 1914 had twice ended in fiasco. So Falkenhayn concluded that Germany itself must take the lead. Conrad was offered eleven divisions under Mackensen to assist. Nominally these would be under joint control, but it was

The Balkans I 1914–18

German attacks	Russian retreat
Austro-Hungarian attacks	Allied attack
Austro-Hungarian retreat	Allied retreat
Serbian counter-attack	Turkish counter-attack
Serbian retreat	German front line
Bulgarian attacks	Austro-Hungarian front line
Romanian attacks	Bulgarian front line
Romanian retreat	Romanian front line

1 Austrian invasion of Serbia repulsed 29 July – 15 December 1914

2 Germans advance up the Morava valley October 1915

3 Allied attempt to take Gallipoli peninsula fails Feb–Dec 1915

4 Bulgarian attack breaks through Serbian formations October 1915

5 Serbian retreat November 1915

6 Romanian forces invade Transylvania 27 August 1916

7 German counter-offensive forces Romanians to retreat Sept–Dec 1916

8 Bulgarian advance forces back Russian–Romanian defence Oct 1916

The Balkans II
September–November 1918

British advance and front line
French advance and front line
Serbian advance and front line
Italian advance and front line
Greek front line

1 Allied front lines, 15 September 1918
2 Allied front lines, 29 September 1918

THE BALKANS

The overrunning of Serbia in 1915 by German, Austro-Hungarian and Bulgarian forces; Romania's ill-fated intervention against the Central Powers in 1916; and the Allies' tardy offensive against Bulgaria from their enclave in Salonika late in 1918.

plain that the German forces would do most of the attacking and the German Command all the decision-making.

Germany also took the lead diplomatically. To aid their offensive and to open a supply route to Turkey, Falkenhayn tempted Bulgaria into the war with an offer of territory it had lost to Serbia in the Second Balkan War of 1913.

Mackensen's plan was to entrap the Serbian forces between three pincers – one Austro-Hungarian and one German in the north, and a Bulgarian in the

south. Operations began on 7 October. Despite their poor equipment, the Serbs, aided by an intimate knowledge of the ground, resisted stubbornly. The weather hampered German supply efforts; Conrad's forces failed to keep pace; and the Bulgarians – no better equipped than the Serbs – gained ground only slowly. As a result there was no encirclement. Nevertheless, the overrunning of Serbia could not be long delayed. The Serb forces were driven steadily back, until, after an epic journey of endurance and suffering over the mountains of Albania, their depleted

War in the Balkans: Bulgarian forces advancing in the region of Monastir.

Painting by Oska Laske (1874–1951) of the storming of Belgrade on 7 October 1915: an imaginative depiction.

army reached the Adriatic. From there it was evacuated by the naval forces of the Entente to the Greek island of Corfu.

Meanwhile, acting on an invitation from the Greek prime minister, a hastily assembled Franco-British force intended to aid the Serbs began landing on Greek territory at Salonika on 5 October. Nothing came of this. By the time the Allied troops were in a position to move forward, the Bulgarians had occupied the mountain chain barring their way and the Serbs had been defeated. Logic dictated withdrawal, especially as a political upheaval in Greece had removed the pro-Allied government of Eleutherios Venizelos and substituted a resolutely neutralist regime. The British were certainly ready to depart. But the French resisted. Joffre wanted to keep the commander of the Salonika expedition, General Maurice Sarrail (who he saw as a threat to his own position), as far away from France as possible, and anyway France had strong interests in this part of the world. As a result, the expedition remained where it was and was even reinforced with the transfer of the Serbian army from Corfu.

The Allies would benefit little from these improvisations, either in late 1915 or during the next two years. They confronted an entrenched enemy securely in control of the high ground, which could not be assailed effectively because of the lack of heavy artillery. With reason Falkenhayn loftily dismissed the Allied emplacement at Salonika as the largest internment camp in the world. His judgement would hold true until the closing weeks of the war.

ANGLO-FRENCH DECISION-MAKING

One large aspect of the war in 1915 remains to be dealt with: the great exertions of the armies of France, with support from the BEF, on the Western Front. Irrespective of the decisions concerning strategy taken by the German Command, these Anglo-French endeavours would occupy much of Germany's attention in 1915 and a considerable proportion of Germany's armed forces.

The decision about where and how France and Britain should act in 1915 would be governed both by the objective realities of the military situation and by the responses to these realities of the political and military leaders of the two nations – particularly, the French High Command.

The facts which dominated all decisions have already

been indicated. Nearly all Belgium and a significant part of France lay under the heel of the conqueror. Correspondingly, the French capital and the Channel ports were perilously close to Germany's grasp. The imperatives to counter this danger and liberate the conquered areas appeared overwhelming. And should the Western Allies be tempted to disregard these matters and divert a significant proportion of their armies to other theatres, the grounds for resisting such temptations were all too evident. France and Britain would be offering Germany the opportunity to unleash another

massive assault in the west, this time against a weakened Allied line, and thereby to fulfil Schlieffen's uncompleted agenda.

Serb forces, retreating to the Adriatic via Albania, seen here on the Vojussa river.

   The French commander-in-chief (General Joffre) and his aides had no interest in looking beyond their own borders. Their chosen method of proceeding was to drive the invader back the way he had come, by massed assaults designed to do better in 1915 what had been done inadequately in 1914. Only in Britain were weighty voices to be heard questioning this course, and then not in military but in

'The war's largest internment camp.' The port of Salonika, where Anglo-French forces landed late in 1915 in an abortive attempt to assist the Serbs. Rebuffed by Bulgarian forces, the Allied troops remained largely immobile until the last weeks of the war.

ABOVE: *David Lloyd George, a consistently energetic prosecutor of the conflict, argued powerfully for a strategy directed away from the Western Front towards the Balkans.*

BELOW: *Winston Churchill, First Lord of the Admiralty in the first year of the war, agreed with Lloyd George, but wanted to attack Turkey rather than the Balkans.*

political circles. Two of the voices were of particular note. David Lloyd George (Chancellor of the Exchequer and soon to become Britain's first Minister of Munitions) and Winston Churchill (First Lord of the Admiralty) saw what the battles of 1914 in the west had revealed about the power of the defensive. They expressed dismay at the prospect of launching further great attacks there.

The concern of Lloyd George and Churchill about the likely human cost of Western Front battles was entirely to their credit. The same cannot be said of the alternative operations which they proposed. Instead of urging only limited attempts at forward moves in the west, they cast around for different areas of operation. So Lloyd George advocated the employment of Britain's forces not against Germany but against one of its allies, either in the Balkans or (later in the war) in the Middle East – what he called, in a singularly inappropriate metaphor, knocking away the props from under Germany. And he demanded to know why Britain's freedom of action should be surrendered to the French commander.

There was no substance in any of this. The prospect of prosecuting the war in France and Flanders may have appeared truly disagreeable. But there was simply nowhere else where Britain's overriding purpose of safeguarding Western Europe from German hegemony, and of liberating the occupied territories could meaningfully be pursued. And anyway, in 1915 Britain had neither the volume of trained soldiers nor the weaponry to conduct a major independent campaign. Its armies could only participate in large operations as an adjunct of the French, who alone possessed sufficient forces in the west to hold back the Germans and perhaps push them out. That certainly entitled Joffre – notwithstanding Lloyd George's displeasure – to a large voice in determining how Anglo-French forces should be employed in 1915.

TRAVAIL ON THE WESTERN FRONT

None of this meant that Joffre, while acting in the only appropriate theatre of war, had now devised appropriate methods of proceeding there: that he had discovered an answer to the conundrum of how to get his forces forward against a well-entrenched adversary possessing all the advantages of defensive weaponry. The events of 1915 made it clear that the answer still escaped him. The dominant features of his campaigns that year were two-fold. The first was the great endeavours of his forces to break through the enemy's front and reach their lines of communication. The second was the conclusive failure of these endeavours.

THE WESTERN FRONT, 1915

The map below indicates the main Anglo-French offensives in the spring and autumn, as well as the German gas attack at Ypres in April. The German attack at Ypres was a minor matter designed to test the use of poison gas. The British attacks at Neuve Chapelle and Loos were also small in scale. The only major offensives of 1915 on the Western Front were launched by the French. Despite prodigious casualties they gained no more ground than the British or the Germans.

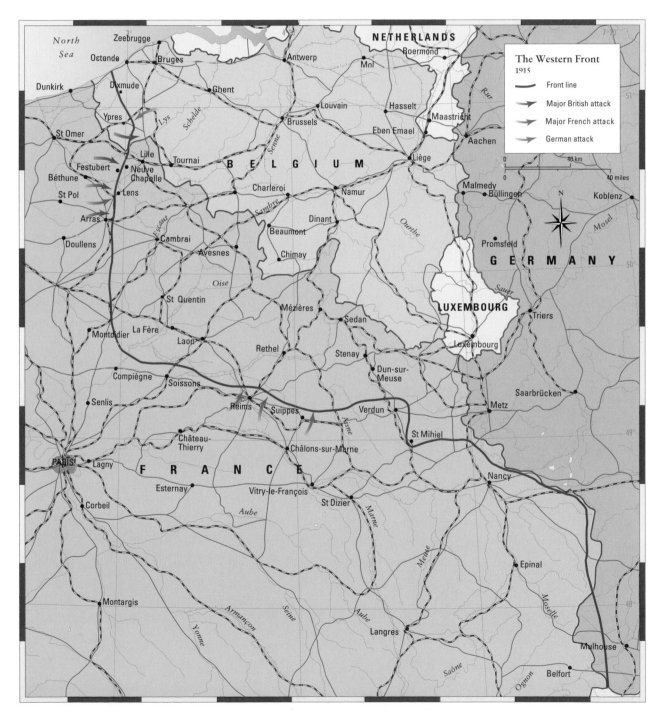

There were some operations on the Western Front in the early part of 1915 which did not involve the French. They consisted of actions first by the British and then by the Germans. Neither was a major operation, certainly when compared with what Joffre would soon be conducting. But each possessed peculiarities rendering it worthy of mention.

On 10 March the British launched their first independent attack on the Western Front. It had been planned as part of a joint action with the French, but when the latter withdrew from it the British decided to go ahead anyway. Designed as an attempt to overrun Aubers Ridge, it managed in its early stages an advance as far as the village of Neuve Chapelle. This limited success resulted from a preliminary bombardment which had been precisely calculated in terms of the quantity of shell required to demolish the German front line. Thereafter, in the absence of further substantial artillery assistance, the attack achieved nothing more.

In itself, Neuve Chapelle was a trivial operation. Yet it had the potential to deliver an important lesson to the British Command. It suggested what could be achieved by an attack preceded by a thorough estimate of artillery requirements. It also suggested the limitations on what that achievement would be. Whether the British Command would be eager to learn either lesson only subsequent events would reveal.

The Germans, it has been pointed out, were primarily remaining on the defensive in the west in 1915. But in April there was a sort of exception, although not a major one. All the armies engaged on this front had been grappling with the problem of how to overcome the dominance of the defensive. One innovation on which the Germans placed considerable hopes, particularly as they believed that their opponents lacked the wherewithal to emulate it speedily, was poison gas. So in a small episode on the Western Front (following an abortive attempt against the Russians which its intended victims had not even noticed) the Germans employed gas in an experimental operation.

At Ypres on 22 April and on the following days, a cloud of airborne poison was directed first against French soldiers and then against Canadians. As the Germans lacked any quantity of shells with which to deliver it, they transported the gas to the front line in cylinders and released it when a favourable wind presented itself. For good reason no reserves had been assembled to exploit the result. It was feared their presence would warn the enemy that something was afoot and so bring down a bombardment on the Germans' lines that might fracture the cylinders and gas their own troops.

At the outset all went well in the use of this horrific weapon. When the gas was released, the French troops, denied the ability to breathe, panicked and fled. But in short order Allied soldiers found that they were able to protect themselves to a sufficient extent by the use of improvised gas masks (which grew more sophisticated as the war proceeded). So the Anglo-French front in the Ypres salient was maintained with only a small loss of territory. British spokesmen

A French grenadier with a mask to protect against gas attack. Poison gas was first used on a large scale in the experimental German action at Ypres in April 1915. Despite its novelty as a weapon, effective gas masks were soon developed to counter its effects.

made much of the Germans' barbarity in employing such a weapon. That would not long delay them in using it themselves.

Up to the end of April, action on the Western Front, be it by the French or British or Germans, was on a limited scale and had availed little. But Joffre was now ready to attempt something much larger. The battle front he selected was in Artois, in the sector ranging from Vimy Ridge to Arras. If the ridge itself could be captured, Joffre reasoned, cavalry could be unleashed into the plain of Douai, the vital rail system seized, and the whole German line unhinged. The British would assist by attacking to the north of the French. Over 1,000 guns, 300 of them heavy, and thirteen divisions were concentrated on the 20-mile front of attack. The French had some reason to be optimistic, at least about an initial success. Because of the Austrian emergency the Germans in Artois were thinly stretched. Opposing the attack they had just four divisions holding two lines of trench of no great strength.

The bombardment began on 9 May and lasted until the 16th. At least in intent, it demonstrated increased sophistication over previous efforts. Heavy guns were directed against both trenches and strongpoints, and against German batteries behind the slope of the ridge. And a mass trench mortar bombardment was unleashed against the German front line. In the centre of the attack these methods proved effective. Despite the protracted nature of the bombardment, the Germans were caught unprepared. A French–Moroccan division even managed to reach the summit of Vimy Ridge. But success was short-lived. The closest French reserves to the Vimy sector were 7 miles distant, and German artillery, largely untouched despite the efforts of the French gunners, soon inflicted such heavy casualties on the Moroccans that they were driven back. That was the total of Joffre's success. South of Vimy, where the German defences were stronger and the bombardment less accurate, no progress was made. And on the flanks (which included a further British effort to secure Aubers Ridge) failure was complete.

Despite these meagre achievements, Joffre derived hope from his momentary advance towards Vimy Ridge. Further attacks were launched throughout May, with little or nothing to show for them. Joffre regrouped. A major offensive was timed for mid June, employing fresh divisions. It was met with an intense German barrage which, at all but one point, stopped the French in their tracks. The exception was in the area of the Moroccan division, which again took Vimy Ridge and again was driven back owing to lack of support. This failure ended Joffre's spring offensive. It had cost him 100,000 casualties, the Germans 60,000.

There was much soul-searching on the part of the French after the battle. It seemed that if only the reserves had been placed further forward to support the Moroccans, a large success might have followed. But of this there is no certainty. The absence of reserves had facilitated the initial success by providing an element of surprise. In addition, as long as the French forces on either side of the Moroccans failed to get forward, Vimy could not have been held for long against unremitting pressure from the flanks.

French soldiers fighting for a plateau in the Champagne during the offensives of 1915. Lithograph by Leon Groc.

Notwithstanding this succession of failures, Joffre remained confident. More heavy guns and reserves placed well forward to seize any opportunity, along with the continuing German manpower crisis on the Western Front, might yet accord him a victory. His view, however, was not universally shared. Allied munitions experts, meeting in June, doubted if sufficient shells and guns could be produced by the autumn to support a renewed offensive. These doubts, combined with the reluctance of Sir John French, the British c-in-c, to repeat the experience of Aubers Ridge, cast doubt on Joffre's design. It was the ongoing Russian emergency that swung opinion back in Joffre's favour. Giving his reluctant consent, the British Minister of War, Lord Kitchener, solemnly observed, 'we have to make war as we must, not as we would like'.

For his attack in the autumn Joffre chose simultaneous offensives in Artois and the Champagne, the goal being the German communications system 50 miles behind the front. This time the major effort would be made in the Champagne. There Joffre assembled eighteen divisions on a 20-mile front and backed them with 700 heavy guns, one-quarter of all French heavy weapons on the Western Front. In Artois the attack would be made by eleven French and five British divisions supported by 420 heavy guns. Close behind both fronts were massed reserves and the ever-expectant cavalry.

In numerical terms the Germans appeared to be in some danger. On the Champagne front they could muster only seven divisions, in Artois just six. However, German trench defences were much stronger now than in the spring. Behind the front line in most areas lay a second line, some 2,500–3,000 yards back and situated where possible on a reverse slope, out of direct enemy artillery observation. It remained to be seen whether these more effective defences would prove sufficient to offset the Germans' relative paucity of front-line troops.

In the Champagne, certainly, French forces made some initial gains. In the centre the concentration of heavy artillery cut the wire and destroyed the German first line, which was then overrun by French troops. They closed next on the second line on a 5-mile front but were unable to capture it. Lack of artillery support, along with the appearance of three fresh German divisions rushed forward by a thoroughly alarmed Falkenhayn, brought the French attack to a halt. A further four days of battering by the French availed them nothing.

In Artois the French aspect of the offensive failed from the outset. Inadequate artillery preparation caused Joffre's troops to encounter uncut barbed wire and intact machine-guns. The only success in this sector was achieved by the British, under the command of Sir John French. To compensate for his insufficiency of artillery, he resorted to the use of poison gas combined with smoke. This proved a mixed blessing. The gas managed to incapacitate more of the attacking troops than those of the enemy, but in some areas the combination of gas and smoke provided concealment for the British infantry during the crossing of no-man's-land. This, together with the movement of German reserves away from the British area to the French sector, enabled Sir John French's forces to capture the first line of enemy trenches on a 4-mile front, along with the village of Loos. Nowhere, however, did the German second line come under threat, and these modest gains were thoroughly overshadowed by the massacre on the following day of two British reserve divisions flung against an alerted and practically untouched German defence. This act of folly cost the two divisions 8,000 casualties and Sir John French his job. He was replaced shortly afterwards by Sir Douglas Haig, commander until then of the First Army.

The rewards of what was to prove Joffre's culminating endeavour of independent offensive warfare were indeed meagre. Total Allied casualties for the

September offensive were almost 200,000, German casualties only 85,000. All that the French and the BEF between them had achieved for this high price was the capture of 12–14 miles of German front line to a depth of about 2,000 yards. This inconvenienced the Germans little. They converted their second defensive system into a new front line and proceeded to construct another series of trenches 3,000 yards behind. The Allies could not influence this process. At the end of a year when they had been repeatedly on the attack, their offensive capacity was for the moment exhausted.

THE LIMITS OF ACCOMPLISHMENT

To all appearances, 1915 seemed a year of very considerable success for the Central Powers and steady disappointment for their opponents.

The one apparent setback sustained by Germany and Austria–Hungary was the entry of Italy into the war on the side of the Entente. But this had not produced the peril to the Habsburg empire which had been anticipated. On the Western Front Germany had held at bay, and had inflicted heavy casualties upon, the armies of Britain and France. On the Eastern Front Russia had been subjected to massive defeats and been deprived of a great amount of territory, including most of Russian Poland. In the Balkans the Central Powers had acquired Bulgaria as an ally and eliminated Serbia as an enemy. Attempts by the Entente to interfere in this last process had accomplished nothing apart from a purposeless diversion of forces from more important fronts.

By contrast, Britain and France had little to show for the year. At much heavier loss to themselves than to their opponents, their offensives on the Western Front had availed nothing. Certainly, the Austro-Hungarian empire continued to look increasingly precarious, but nowhere had it proved possible to exploit this weakness.

Nevertheless, the year had not served to further Germany's main purpose. Its major operations had been designed not to accomplish the defeat of its principal adversaries, Britain and France, but to prop up its failing ally Austria–Hungary against the Russians and Serbs. When the year was over, the major challenge facing Germany at the outset was facing it still: the mounting mobilization of men and industry in France and above all in Britain and its dominions, and the continued British command of the sea, carrying with it the ability to draw on the industrial and agricultural resources of neutral powers (most notably the United States) for the Allies' war requirements.

Sir John French, British commander-in-chief in France and Flanders 1914–15. His succession of unavailing offensive actions in 1915 caused him to lose his post at the end of the year.

CHAPTER THREE

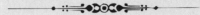

PERIPHERIES

TURKISH DIGNITARIES *arranging the surrender of Jerusalem to the British, December 1917. The capture of the city was hailed by Lloyd George as a triumph. In reality it was insignificant compared with the blood letting of the Third Ypres campaign over which he had just presided.*

PERIPHERIES

THE EXPANDING CONFLICT

The deployment of the German army on the territory of France and Belgium in the west and Russia in the east fixed the locations of the main battlefronts. Nevertheless, there would be other campaigns conducted in this war hundreds or even thousands of miles from these fronts. There were the battles already mentioned which Austria–Hungary fought against Italy, and which Bulgaria fought against a combined Allied expedition in Salonika. There were also a number of far-flung campaigns; for example, those which the Allies conducted against the Turkish empire, and those which Britain (along with, in one instance, Japan) waged against the German colonial empire.

Japan at war. Following Japan's declaration of war against Germany on 23 August 1914, Japanese forces attacked the German naval base of Tsingtao in China. Japanese soldiers are here seen occupying a German battery after the fall of Tsingtao in November 1914.

ACTION IN THE PACIFIC

The first series of campaigns was instituted on 6 August 1914 when the British government sent telegrams to Australia, New Zealand and South Africa asking them to mount expeditions to seize nearby German colonies, and requesting its ally Japan to hunt down German merchant shipping in Far Eastern waters.

Australia and New Zealand were only too willing to comply. Expeditionary forces were hastily organized and soon overwhelmed the isolated and outnumbered German garrisons. By the end of September Samoa, German New Guinea and the islands of the Bismarck Archipelago had been captured. The German empire (such as it was) in the South Pacific had been liquidated at a cost of six Australian dead and fourteen wounded. The New Zealanders lost not a man.

Operations in the northern Pacific began on 12 August 1914 with the destruction by the British China Squadron of the German radio station on the island of Yap. From then on, affairs did not proceed as Britain had intended. Japan had its own ambitions in this area and was not amenable to British direction. Essentially, Japan was eager to enter the war but unwilling to confine its activities to the capture of a few merchant ships. Despite the efforts of the British Foreign Secretary, Sir Edward Grey, to restrain it, Japan declared war on Germany on 23 August and proceeded to mop up Germany's island colonies in the northern Pacific. By October the Marshall, Caroline and Marianas island groups were in Japanese hands.

But Japanese ambitions did not end there. At Tsingtao on the coast of China the Germans possessed a naval base of some strategic importance. It was heavily defended. On the landward side the garrison of 6,000 was supported by trenches, barbed wire, machine-guns and field artillery. The seaward approaches were guarded by fifty-three heavy guns, some of which could be traversed to fire over land. Soon after the declaration of war, Japanese warships instituted a blockade of the base, while the army began assembling an invasion force. By the end of September a contingent of 50,000 men, supplemented by one and a half battalions of British troops rushed up to establish a token presence, was ready to advance. Japanese tactics were hardly subtle. They took the form of a succession of massed frontal assaults. After much hard fighting and only when many of the large German guns had run out of ammunition, Tsingtao fell on 7 November. The Japanese sustained 6,000 casualties, the Germans 700. The fall of Tsingtao saw the demise of the entire German empire in the Pacific.

THE WAR IN AFRICA

The situation in South Africa was more complicated than in other British Dominions. There the government was split between those Afrikaners who had arrived at an accommodation with Britain after the Boer War and those who had not. When Britain's call to invade German South-West Africa (present-day Namibia) arrived, the dissidents rebelled. By January, government forces led by the South African statesmen Louis Botha and Jan Smuts had crushed the revolt. Their attention now turned to the Germans. Two wireless stations on the coast of South-West Africa had been seized as early as 10 August. Reinforcements were now sent to these ports, and together with two columns invading from the south, set off in pursuit of the small German garrison. Outnumbered twenty to one and with little scope for concealment in the plains and deserts, the Germans were forced to capitulate. After a short war which saw few major actions the territory fell to the South Africans on 9 July 1915. South African casualties were fewer than 500.

In West Africa operations were under the control of the Colonial Office. They started well for the Allies. The German colony of Togoland, a narrow strip of territory sandwiched between British and French possessions, was overrun in three weeks by converging attacks.

Jan Christian Smuts, South African soldier and statesman, first suppressed a revolt of pro-German elements in South Africa and then participated in campaigns first in German South-West Africa and then more protractedly in German South-East Africa. He ended the war as a member of the Imperial War Cabinet in London.

Germany's other West African colony, Cameroon, proved more intractable. Here the British and French found a well-led German force supplemented by native troops, equipped with numerous machine-guns and light artillery, operating in mountainous jungle country. The resourceful German commander, Colonel Zimmermann, was in fact never defeated. For eighteen months he eluded or occasionally beat his pursuers. In February 1916, when he was finally cornered, he interned his force by crossing into neutral Spanish territory.

In German South-East Africa, the British, against all expectations, faced what proved to be the longest campaign of the war. The German commander, General Paul von Lettow-Vorbeck, after inflicting two defeats on British troops invading from Kenya, abandoned conventional warfare and proved himself a guerrilla commander of genius. For the next four years he led various British commanders, including Smuts (himself no stranger to guerrilla war), on a fruitless hunt through the colony. The keys to his success were his superior mobility and his quick realization that indigenous troops, being acclimatized, could be utilized to good effect. Finally, in 1918, the overwhelming numbers

employed against him drove his force from South-East Africa. Lettow's riposte was to invade neighbouring Mozambique, elude his pursuers and re-enter the German colony. From there he launched an invasion of Northern Rhodesia (present-day Zimbabwe). Only the end of the war in Europe brought his operations to a halt. He surrendered on British territory on 25 November 1918, the last German commander to do so.

GALLIPOLI

Turkey's entry into the war on 29 October 1914 on the German side opened another theatre of war. The principal reason for Turkey's decision was its long-standing rivalry with Russia, one feature of which was Russia's ambition to obtain a warm-water port and access to the Mediterranean. During most of the nineteenth century, Turkey had looked to Britain and France to shore up its position against Russian expansion. And a remnant of this orientation remained in 1914 in the person of a British admiral, Sir Arthur Limpus, chief naval adviser to the Turkish government. But with the emergence of Germany as both the

African Askari forces operating with the Germans in East Africa in 1914. The German authorities recognized more quickly than their adversaries the value of native troops under these conditions.

major military power in Europe and a devoted adversary to Russia, Turkish allegiance turned towards Berlin. So by 1914 Britain had ceased to count in Turkey and a German, General Liman von Sanders, had become the principal military adviser to the Turkish army.

In late October 1914 Turkish warships bombarded Russian ports on the Black Sea, thus precipitating war with the Entente Powers. For the next two years Turks and Russians fought extensive but inconclusive campaigns in the Caucasus.

Despite these operations, it soon became clear that the major Turkish adversary would be Britain. Initially, British actions against the Turks brought upon it a succession of humiliating – if hardly mortal – defeats; but in the long term they had the consequence of destroying the Turkish empire. For the British, this was ironical. A weak Turkish empire on the flank of the route to India had served British interests well. Once this empire was gone, Britain was obliged to adopt a more direct role in areas such as Palestine and Mesopotamia (Iraq) and Persia (Iran), in order to secure the trade routes of its own imperial position. The process would, in the years after 1919, prove somewhat uncomfortable.

At first there seemed no overwhelming imperative for Britain to undertake extensive campaigns against Turkey. Two vulnerable points adjacent to Turkish territory and of great moment to Britain – the Suez Canal and the Abadan oil refinery – were soon secured: the first by beating off a hastily planned Turkish attack across the Sinai, the second by landing a small contingent of troops from the Indian army at Basra on the Persian Gulf. No one in 1914 imagined that these small affairs guarding the trade routes and oil supplies of the empire would lead to operations that ultimately involved one million Allied troops.

What, however, would prove to be the first major operation by British forces against the Turks occurred neither in Egypt nor in the Persian Gulf. Instead it was directed towards reopening trade relations with Russia through the Black Sea and simultaneously driving Turkey out of the war.

The origins of this operation can be found in the belief of the First Lord of the Admiralty, Winston Churchill, that he could use a purely naval force substantially to alter the course of the war on land. In November and December 1914, he cast about for an operation in German waters or in the Baltic which might achieve this result. All were vetoed by his admirals on the sensible ground that the risk to the Grand Fleet in confined waters infested by mines, submarines and torpedo-boats was too great. Thwarted in the north, Churchill looked further afield. Here he was aided by a particular conjunction of events. The Russians, hard pressed by the Turks in the Caucasus, had asked Kitchener for some kind of action against Turkey to relieve this pressure. Churchill now devised a scheme to which he secured a halting assent from an admiral of no great repute (Sir Sackville Carden) who happened to be commanding a squadron lying off the Dardanelles. The plan involved forcing the Dardanelles straits by a fleet of old (and therefore superfluous) battleships which would then proceed to Constantinople, overawe the Turks and oblige them to withdraw from the war.

The effect of all this, Churchill argued, would be wide-ranging. Among other things, the still-neutral Balkan states would be so impressed that they would rally to the Allied side and form a great phalanx of British-led forces proceeding along the Danube valley. Thereby Austria–Hungary and Germany would be threatened from the south-east.

At least momentarily this scheme appeared to make sense to Britain's political leadership. No vital ships would be risked and apparently no troops would be needed. And if it did not work out, the operation could (as Kitchener stressed at the outset, but soon forgot) be called off with nothing lost.

So Churchill received permission to commence the naval attack. The ships were assembled and fire was opened on the outer forts of the Dardanelles on 19 February 1915. Although many Turkish guns were knocked out, this was by the action of landing parties rather than by the ships' guns. Perhaps this was not made clear to the London authorities.

Herbert Horatio Kitchener, Britain's most famous soldier at the outbreak of war and swiftly brought into the government as Minister of War. Although predominantly a supporter of the western strategy, he allowed himself to be drawn into agreeing to the naval campaign at the Dardanelles and then the dispatch of troops to Gallipoli.

From then, on naval operations proved unproductive. Inside the straits the Turkish defences consisted of large guns in the forts at the Narrows, minefields and mobile batteries of howitzers along the shore lines. The fleet could not demolish the guns at the Narrows by long-range fire, or clear the mines protected by the howitzers, or land naval parties to undertake the work of demolition. It could not even protect itself. A concerted attack on 18 March, designed to destroy the forts and allow small craft to sweep the mines, failed in both these tasks and suffered prohibitive loss: one-third of the naval force intended to overawe Constantinople into surrender was either sunk or put out of action in what was still only a preliminary episode.

Given that the British War Council had only authorized a naval operation at the Dardanelles, and that this had manifestly failed, the time had apparently arrived to follow Kitchener's advice and call off the affair. Yet this wisdom seems to have occurred to no one – least of all Kitchener himself, who now opined that if the navy could not get through on its own then a military force must be sent to the Dardanelles to occupy the Gallipoli Peninsula, suppress the forts, and allow the fleet to resume its endeavours. It so happened that 70,000 troops were in the

GALLIPOLI, 1915

The attempted naval forcing of the Dardanelles.

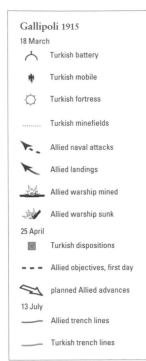

Gallipoli 1915

18 March

⌒ Turkish battery

⋔ Turkish mobile

☼ Turkish fortress

.... Turkish minefields

➤ Allied naval attacks

➤ Allied landings

Allied warship mined

Allied warship sunk

25 April

▪ Turkish dispositions

– – – Allied objectives, first day

➪ planned Allied advances

13 July

—— Allied trench lines

—— Turkish trench lines

near vicinity of the Dardanelles or could be dispatched there: these consisted of a naval division, a Regular Army division (the 29th), an Australian and New Zealand contingent (soon to be known as the Anzacs), and a French division. A putative commander for this force, Sir Ian Hamilton, was rushed to the scene and speedily concluded that the naval operation would only succeed if a military landing was first carried out. It was to take place on 25 April.

The landing was divided into two sections. At Cape Helles on the toe of the peninsula, the 29th Division had the task of advancing on the Kilid Bahr Plateau and silencing the forts at the Narrows. Further north, Anzac forces were to land from the sea, cross the peninsula and so prevent Turkish reinforcements from moving south towards the 29th Division. The French, meanwhile, were to effect a diversionary landing on the Asiatic shore near Kum Kale.

At the toe of the peninsula, which was the vital part of the operation, very little proceeded as planned. The flanking forces landed unopposed, but the main contingent encountered underwater obstructions which caused them to disembark in the sea and brought them under unrelenting fire from Turkish riflemen and machine-gunners whom a bombardment from the ships had left unscathed. By nightfall a precarious hold had been established on the beaches, but at such a cost that no early advance inland could follow.

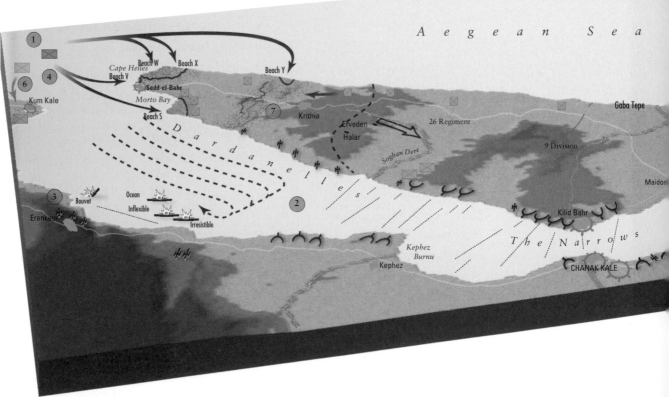

The River Clyde *at the landing of British troops at Cape Helles, 25 April 1915. The sally ports cut in the side of the ship, intended to facilitate the egress of troops, instead served as a focal point for Turkish machine-guns.*

RIGHT: *Kemal Atatürk, seen here as leader of Turkey in the inter-war years, first came to prominence as commander of Turkish forces resisting the Anzac operations on the Gallipoli Peninsula.*

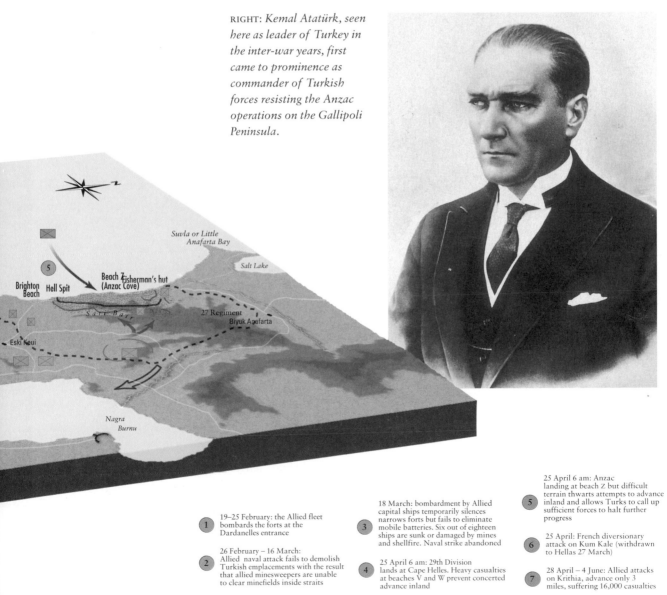

1. 19–25 February: the Allied fleet bombards the forts at the Dardanelles entrance

2. 26 February – 16 March: Allied naval attack fails to demolish Turkish emplacements with the result that allied minesweepers are unable to clear minefields inside straits

3. 18 March: bombardment by Allied capital ships temporarily silences narrows forts but fails to eliminate mobile batteries. Six out of eighteen ships are sunk or damaged by mines and shellfire. Naval strike abandoned

4. 25 April 6 am: 29th Division lands at Cape Helles. Heavy casualties at beaches V and W prevent concerted advance inland

5. 25 April 6 am: Anzac landing at beach Z but difficult terrain thwarts attempts to advance inland and allows Turks to call up sufficient forces to halt further progress

6. 25 April: French diversionary attack on Kum Kale (withdrawn to Hellas 27 March)

7. 28 April – 4 June: Allied attacks on Krithia, advance only 3 miles, suffering 16,000 casualties

The battle for Anzac Cove. Australian reinforcements arriving on the peninsula.

Further north the plan had also gone awry. The small boats bringing the Anzacs from the warships lost direction in the dark and landed them 1 mile to the north of the intended beach, facing sharply sloping ground cut up by tortuous ravines. Turkish fire was heavy enough to inflict many casualties, especially among junior officers. These losses, together with the confusion about the landing place and the inadequacy of such maps as had been provided, prevented a concerted drive inland. By the time some order was established, Turkish opposition had stiffened. At nightfall, so far from advancing, the Anzacs were fortunate to have any hold on the peninsula at all. During the days that followed, reinforcements flowed in, with the result that some small advances were made and a defensive perimeter established. But although this proved strong enough to fling back with great slaughter a concerted Turkish attack on 19 May, it presented no base for a further advance across the peninsula with its imposing terrain and well prepared opposition.

The precarious landings achieved at Helles and Anzac on 25 April and the small advances made on the following days would prove almost the full extent of Allied success on the Gallipoli Peninsula. From late April until early June the British and French forces at Helles made repeated attempts to redeem this situation by breaking through an increasingly sophisticated Turkish defence. These actions, which were characterized by unimaginative frontal assaults supported by inadequate amounts of firepower, failed invariably. Trench warfare of the kind so familiar on the Western Front set in.

A last endeavour to break the stalemate was made in the northern sector in August. The Anzac forces attempted to capture the Turkish positions on the dominant ridge as a first step to advancing across the peninsula. At the same time a new force was landed to the left of the main operation to capture flat ground around Suvla Bay. Its purpose was to create a base from which the entire northern force could be supplied.

The plan proved too ambitious. The debilitated Anzac forces lacked the necessary hours of darkness required to reach the high ground ahead of alerted Turkish reinforcements. At Suvla a base was eventually established, but this

contributed nothing to the main operation. The setback dispelled all hope of success. Eventually the authorities in London took the only logical decision on offer and agreed to withdraw the Allied forces. Anzac Cove was evacuated on 19–20 December 1915 and Helles in January 1916, on each occasion with scarcely the loss of a single Allied soldier.

In the course of the campaign the Allies suffered 180,000 casualties, the Turks more than twice as many. What became the best-known sideshow of the war was now at an end.

The Gallipoli operation has often been described as the great lost opportunity of the war. The foregoing account provides no evidence that it was anything of the kind. The forces landed on the Gallipoli Peninsula never managed to get within miles of their objectives. And even if they had succeeded in overrunning the peninsula and suppressing the Turkish forts, it stretches

Anzac Cove, scene of the landing by Australian and New Zealand forces on 25 April. Taken late in the campaign, this photograph reveals the dug-outs in the hills, the pier constructed (under enemy fire) to deep water, and soldiers bathing. But most of all, it makes clear the highly defensible nature of the Gallipoli terrain.

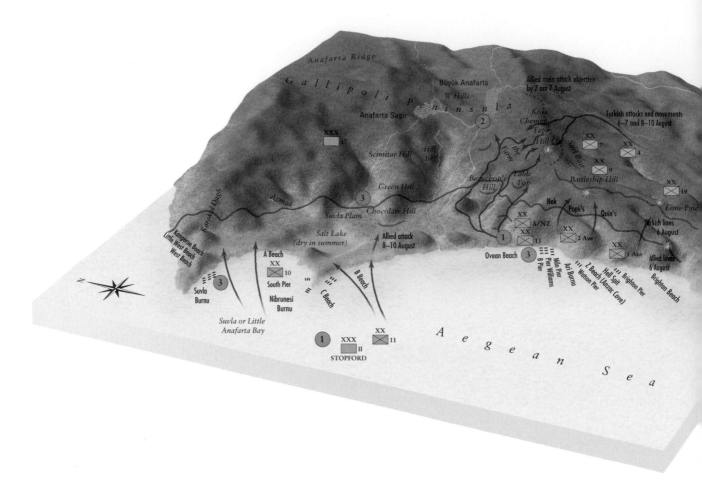

THE GALLIPOLI PENINSULA:
SUVLA BAY

*The last series of attempts
to overrun the Gallipoli
Peninsula in August 1915.
Once again, the terrain and
an alerted Turkish defence
proved too strong for the
Anzac troops.*

credibility to argue that a few battleships approaching Constantinople would
have terrified the Turks into capitulating. Nor, it needs to be added, was the
presence of Turkey in the war of any real consequence to the main conflict being
fought in Europe.

MESOPOTAMIA AND PALESTINE

Having had one major success against Britain in 1915, the Turks would have
another, but more transitory, the following year. This was in the unlikely region
of Mesopotamia.

As mentioned earlier, back in 1914 the British, employing Indian troops, had
secured the oil refinery at Abadan by landing forces at Basra in Turkish-held
Mesopotamia. This gave the Viceroy of India (Lord Hardinge) and his
commanders on the spot a stake in operations. They viewed the toehold in the
Turkish empire as an excellent opportunity to demonstrate to Muslims abroad
(and at home) the power of British India. In London the War Council tended
towards caution, but in this matter they were prepared to allow the authorities in
India to make the running.

Essentially, operations in Mesopotamia over the next three years consisted of
advances along the two major rivers (Tigris and Euphrates) to prevent the Turks

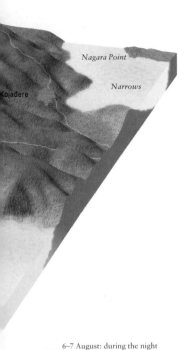

Nagara Point

Narrows

Kojadere

1 6–7 August: during the night the Anzacs launch attacks toward Sari Bair and Lone Pine. Simultaneous landings are made at Suvla Bay

2 8 August: Turkish reinforcements are rushed forward on time to catch the landing force in the open plain. They are driven back to their start lines

3 August: after savage fighting on the Anzac front the situation becomes deadlocked. The British/Anzac force is evacuated during the night of 18–19 December

from counter-attacking whatever happened to be the last point seized. So Abadan had to be protected by driving the Turks from Basra, Basra by driving them from Amarah, Amarah by seizing Kut. What fuelled these forward movements was a determination by the British commanders on the spot and the viceroy in India ultimately to capture the prize of Baghdad.

Initially this policy was spectacularly successful. General Sir John Nixon and his subordinate Major General Sir Charles Townshend advanced hundreds of miles upriver from Abadan and by September 1915 had seized Kut. The Turks had fled before this small force, which had used such unorthodox implements as canoe-borne infantry and shallow-draft river gunboats. The second phase was just as spectacularly unsuccessful. An advance from Kut overextended the British forces, which then encountered groups of German-reorganized Turkish formations. Townshend was driven back to Kut. A siege ensued. All efforts to break it failed and Townshend with 13,500 troops was forced to surrender on 29 April 1916.

Although the moment seemed ripe, no thought was given to retreating to Amarah or Basra or wherever it was considered that a defence of Abadan could be sustained. Instead there were declarations about restoring British prestige in the east. A new commander (Lieutenant General Sir F. S. Maude) was sent out and three extra British divisions despatched.

Under Maude, affairs took on an altogether more methodical aspect. Advances were now made over short distances which could be covered by the field artillery. By a series of these deliberate actions Kut was retaken by the British on 24 February 1917. Turkish resistance thereupon crumbled. Two weeks later British troops entered Baghdad. Any further action in this region was now without purpose. Nevertheless, advances against the Turks continued until an armistice was signed in October 1918.

All told, the campaign cost the British some 90,000 casualties (including Maude, who succumbed to cholera). Turkish losses were probably twice as great.

At the same time a further campaign against the Turks was being played out in Sinai and Palestine. In February 1916 General Sir Archibald Murray (c-in-c Egypt) announced that the Suez Canal could best be defended from Turkish intrusion by occupying the Sinai Peninsula. In London the War Council half-heartedly agreed. This was all Murray needed. He pushed forward and defeated the Turks at Romani in August. He then paused while a water pipeline and railway track were built across the desert to supply his forces on the far side of the Sinai. Murray could have stopped here but his ambitions were growing. He determined to seize Gaza and drive the Turks out of Palestine. Stubborn Turkish resistance thwarted his attempt, costing the British 6,500 casualties and Murray his job.

Notwithstanding this rebuff, operations in Palestine would continue. The new British Prime Minister (Lloyd George) was much taken with peripheral

warfare. He had just been denied a campaign in Italy, and he now determined to have one in Palestine. Murray was replaced with General Sir Edmund Allenby. His force was increased to seven British infantry divisions (this constituted the most severe drain of British manpower away from the Western Front of any campaign) and three cavalry divisions. And Lloyd George ordered him to take Jerusalem. This he did. After inflicting defeats on the Turks at Beersheba by dint of a massed cavalry charge (one among many contenders in a long list of last cavalry charges in history), and at Gaza by the use of overwhelming firepower, he finally entered Jerusalem on 9 December 1917.

The sight of Allenby walking bareheaded into the holy city was not the only romantic aspect to this war. Throughout 1917 the Turks were forced to divert troops and resources to the Hejaz where an Arab revolt against their rule was aided by British money, arms and T. E. Lawrence. The effect of Lawrence and his Arab irregulars on the war in the Middle East has been much exaggerated, not least by Lawrence himself. Nevertheless, his force did tie down many more Turkish soldiers than those employed against them and was in this sense cost effective.

After the dramatic events of 1917, the Middle Eastern war in 1918 was something of an anti-climax. The German attack on the Western Front in March drew troops and attention away from the region. Only in the last weeks of the war was serious campaigning resumed. At Megiddo in September 1918, a British offensive employing methods strikingly similar to those on the Western Front – for example, the delivery of 1,000 shells per minute upon the Turkish defences – achieved a crushing victory. This operation was aided by Lawrence, who had been continuing to harry Turkish communications on the desert flank. As the Turks now fell back on Damascus, the Arab irregulars and some Australian cavalry hastened to cut them off. Damascus fell to both elements on 1 October. Turkey's last army was surrounded. Before the end of the month the Turkish empire had capitulated.

The climax in Mesopotamia: British troops enter Baghdad, 11 March 1917.

SUMMING UP THE SIDESHOWS

How are we to assess the utility of these peripheral campaigns? Were the men and munitions and supplies expended on them worthwhile? The Pacific and African campaigns are perhaps the easiest to judge. If the value of the German colonies is not subjected to close scrutiny, it might be thought that 18,000 casualties was a reasonable price to pay to secure them. Moreover, most of the troops that fought in these campaigns were Indian and indigenous forces that were unlikely to have been employed on a major battlefront. Nor were these operations extravagant in their use of shipping and munitions.

The other campaigns were of more dubious worth. Operations against

THE MIDDLE EAST, 1914–18

This map shows the main thrusts by the British against the Turks. By 1918 the Turkish empire in the *Middle East was entirely in the hands of the British, a legacy that was to bring only trouble.*

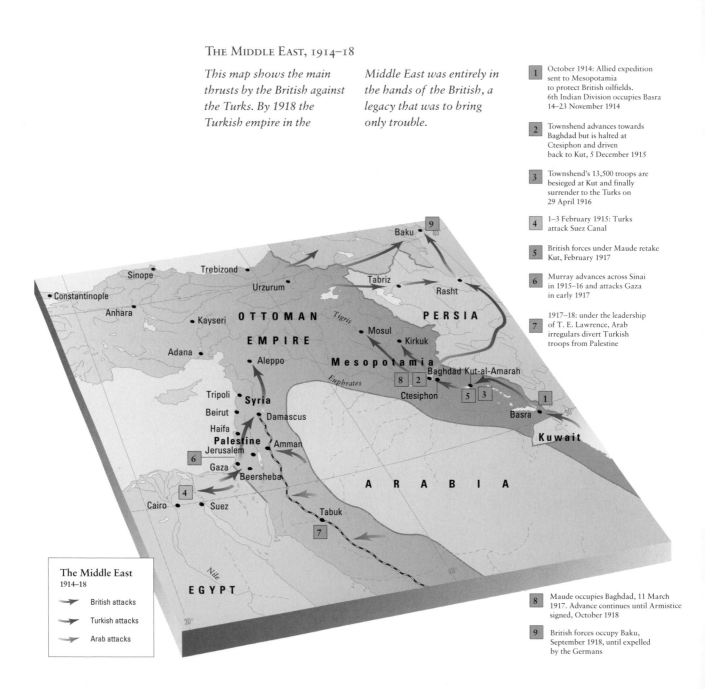

1　October 1914: Allied expedition sent to Mesopotamia to protect British oilfields. 6th Indian Division occupies Basra 14–23 November 1914

2　Townshend advances towards Baghdad but is halted at Ctesiphon and driven back to Kut, 5 December 1915

3　Townshend's 13,500 troops are besieged at Kut and finally surrender to the Turks on 29 April 1916

4　1–3 February 1915: Turks attack Suez Canal

5　British forces under Maude retake Kut, February 1917

6　Murray advances across Sinai in 1915–16 and attacks Gaza in early 1917

7　1917–18: under the leadership of T. E. Lawrence, Arab irregulars divert Turkish troops from Palestine

8　Maude occupies Baghdad, 11 March 1917. Advance continues until Armistice signed, October 1918

9　British forces occupy Baku, September 1918, until expelled by the Germans

The Middle East
1914–18

→ British attacks

→ Turkish attacks

→ Arab attacks

THE PALESTINE CAMPAIGN

Begun by General Murray in 1916, the campaign was brought to a successful conclusion by General Allenby in a series of victories in 1917 and 1918.

1 After two failed attempts to take Gaza in early 1917, General Murray is replaced by General Allenby

2 Leaving only three divisions at Gaza, General Allenby attacks Beersheba, which falls on 31 October

3 Turkish forces counter-attack but by 7 November are beaten back

4 The Desert Mounted Corps head across country towards the coast, forcing the Turks out of Gaza. British occupy the city 7 November

5 British troops capture Jerusalem on 8 December

6 Colonel T. E. Lawrence and his Arab irregulars disrupt the Hejaz railway

7 Further offensives were curtailed in early 1918 as the Western Front needed reinforcements. It was 19 September before the campaign could resume

8 Allied cavalry capture Nazareth on 20 September

9 By 22 September the Turkish Fourth Army is in retreat. Some units surrender near Damascus, the rest near Amman

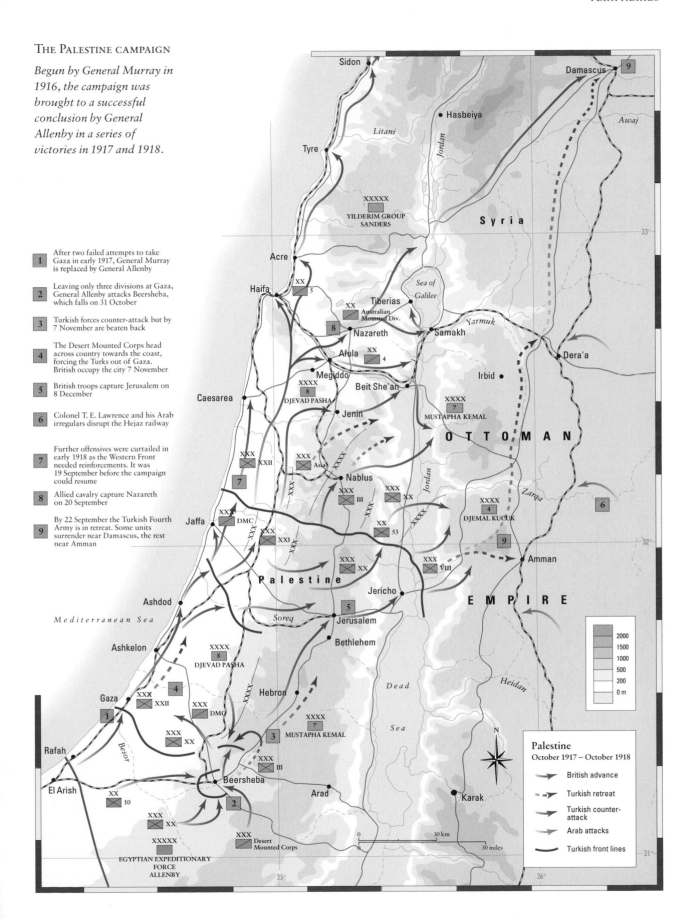

Palestine
October 1917 – October 1918

→ British advance
⇢ Turkish retreat
→ Turkish counter-attack
→ Arab attacks
━ Turkish front lines

T. E. Lawrence of Arabia, in characteristic garb. A British intelligence officer, he became one of the leaders of the Arab revolt against the Turks.

Turkey cost Britain, the Dominions and France well over 250,000 troops, while the main enemy (Germany) suffered only a derisory number of casualties. The British and French forces lost in the sideshows could, in 1918 on the Western Front, have made a significant contribution to Allied survival in the first half of the year and Allied success in the second. A further issue concerns supply. Throughout periods of chronic shortages of munitions and shipping, the Allies expended significant amounts of both on these campaigns. The munitions factor was serious in 1915 and 1916, as was shipping from the time Germany launched its unrestricted submarine campaign in 1917.

Finally, there was the essential purposelessness of most of these undertakings. Only Gallipoli had an identified objective, and that lay in the realm of fantasy. The remainder began with limited and defensive purposes – the security of the Suez Canal, the protection of the Abadan oil refinery – which made a degree of sense. But very soon they developed grandiose objectives which, in addition to their excessive cost, offered only dubious benefits. The acquisition of territories formerly part of the Turkish empire proved a drain on Britain's diminished resources for years to come. Moreover, Britain in the course of the war made contradictory promises to its Arab and Zionist supporters which produced unhappy long-term consequences. In short, for Britain and France, their 'moment in the Middle East' was a legacy they could have well done without.

The last days of the Turkish empire: retreating Turkish forces march through the city of Damascus in late 1918.

CHAPTER FOUR

1916

BATTLE OF THE SOMME, September 1916. The old German front line near Ovillers, now in British hands. The white chalk, thrown up by the digging of trenches, was characteristic of the Somme battlefield. The ruins of a wood can be seen on the horizon. By the end of the campaign the entire Somme battlefield had been reduced to this type of moonscape.

1916

GERMANY FACES WEST

At the end of 1915 the rival high commands had once again to make plans for the year ahead. And, yet again, this process involved the attempt to combine truths evident from the first day of the war with lessons learned painfully since.

For Germany there remained the major difficulty that it was engaged in a war on two fronts. But in an important respect this appeared less of a problem than it had been a year before. Certainly, Austria–Hungary continued in a precarious condition and Russia still constituted a challenge to it. But the savage blows which the German army had rained upon the tsar's forces in the course of 1915 had demonstrated that Russia was no great threat to Germany itself and would require much recuperation before it could again menace the Habsburg empire.

This gave Falkenhayn the opportunity to wage the war to which, all along, he had assigned priority: that on the Western Front against France and Britain. But to accomplish victory there, he must come up with a solution to the great conundrum posed by the battles of 1914 and 1915. The emergence of elaborate trench systems supported by massive quantities of defensive weaponry threatened to thwart any projected German offensive in 1916 as thoroughly as it had done Joffre's great endeavours of the year before.

Sir Douglas Haig, British commander-in-chief on the Western Front (in succession to Sir John French) 1916–18.

Falkenhayn rejected this negative view. (His reasons would simultaneously be embraced by his British and French opposite numbers.) He reasoned that all previous offensives in the west had only failed because of the absence of sufficient high explosive shells – the principal weapon of attack. Now German industrial mobilization had rectified that deficiency. So he would stockpile a huge quantity of shells and direct it against a selected section of the Allied line in the west. Thereby he would create a desert and call it victory.

Falkenhayn's proposed victims were the soldiers of France. In a way, this was odd. He had often identified Britain as the main enemy. To square this circle, he characterized the French army as Britain's best sword. Plainly, this was deception. Whatever had been the case in 1914 and

Zensiert
Paul Hoffmann & Co.
Berlin-Schöneberg.

1954.
phot. Bild-und Film-Amt.

Die Abwehrkämpfe im Westen.
Schweres Geschütz im Feuerkampf bei Deinze.

1915, by early 1916 Britain was training and equipping a mass army which would soon be its own best sword. Perhaps, in choosing not to attack it, Falkenhayn was recalling the bloody rebuff of First Ypres. Perhaps he was calculating that the French, after a year and a half of bloodletting, were at their last gasp and might drag Britain down with them. Most likely, he was just devising a big offensive at a hopeful place and trusting that some sort of positive result would flow from it.

Falkenhayn's chosen target was Verdun. This was an important army administrative centre protected by a great array of fortresses astride the River Meuse. Back in 1914 it had proved a hinge of the French defence in this sector of the line. His intention was to eliminate the salient created by that earlier fighting and to capture Verdun. But thereby he envisaged doing a great thing. By employing unprecedented amounts of high explosive shell, he would first annihilate the soldiers guarding the forts and the city. Thereafter, he would impose insupportable losses on the reinforcements which the French Command would be forced to feed into the battle in a futile attempt to hold on to this symbolic site. (It may be noted that Falkenhayn at one time claimed that he did not even need to capture Verdun in order to cripple France. Just killing French soldiers would be enough. Nothing about the operations he actually commanded, or about the logic of conducting an offensive on the Western Front, supports this view. What he seems to have been doing, and that probably only after the event,

A German heavy gun in action on the Western Front. Increasingly, these weapons became the dominant force in battle in the Great War. The original caption gives the catalogue number of the illustration and indicates that it was censored by the wartime authorities.

was arguing that whether he succeeded or failed he would still be the winner.)

Falkenhayn planned his campaign for mid February. He was taking a chance on the weather at this time of year. But if fate was favourable in that respect, he would be managing to act well in advance of his opponents.

THE ENTENTE MAKES DECISIONS

So dominant was Germany's position among the Central Powers that its High Command could more or less decide on strategy for the coming year without troubling about its allies. Nothing of the sort was true on the other side. The strategy of the Entente Powers must be the product of intense consultation. In the event, Allied decision-making for 1916 gave rise to noteworthy unanimity. The high commands of France, Britain, Russia and Italy proposed, and the political commands decided, that their armies should take the offensive simultaneously. Thereby they would deny to Germany the opportunity of employing its central position to move forces first to one front and then to another.

This meant waiting until the middle of the year. Only then would Britain have trained and equipped its 'Kitchener' armies of wartime volunteers, and Russia have both recovered from its misfortunes of 1915 and mobilized its industry to the extent necessary to provide its army with adequate munitions. At that point

The 'Kitchener' armies in training. These troops would see action on the Somme in the summer of 1916.

British and French forces would attack the invader side by side in the middle of France, while Russia struck at the Germans on their northern front. Simultaneously, tsarist forces would move against the Austro-Hungarians on their south-western front and the Italians would attack them on the Isonzo, thus stretching the Habsburg armies to what was expected to be breaking point.

These operations looked very different from those being devised by

Falkenhayn. But they had one important similarity. Allied plans, like German, were dependent on the view that sufficient quantities of high explosive shell were now being manufactured to terminate the dominance of the defensive. The British might also place some hopes on a new weapon which they were developing in great secrecy – an armoured fighting vehicle capable of crossing trench lines and acting in a small way as a mobile artillery platform. But the tank, as it was codenamed, was to be an accessory only. The instrument of victory would be, as on the German side, prodigious numbers of shells.

SUPPLYING THE ARMIES

For the extensive battles thus envisaged, munitions in huge quantities would be required. How well had the munitions industries in the main combatant countries been geared to the demands of this war?

Initially, the continental powers had entered the war with reasonable amounts of light guns and shrapnel shells. What was lacking were the heavy guns and high explosive shells which it was soon revealed were necessary to batter down trench defences and dug-outs. So in the first year of the war there were various 'shell scandals' in the belligerent countries as this deficiency became manifest.

To begin with it was Germany which was best placed to react. It entered the

The casting of a large gun in a German cannon factory.

Women wearing respirators in a British shell factory. As the war went on women played a larger role in the manufacture of munitions, despite its many hazards.

war with a well-developed munitions industry, and initially did not call up many munitions workers for the armed forces. As one indicator, Germany was able to increase its production of heavy artillery from 38 guns per month in 1915 to 330 per month by the autumn of 1916. Shell production was increased in due proportion. Hence by early 1916 Germany was better equipped with heavy guns and shells than any other power. Whether it had enough for its purposes at Verdun was another matter.

France was slower to respond. In an endeavour to compensate for its smaller population, many workers from heavy industry were called to the colours in 1914. It then proved exceedingly difficult to get the army to disgorge them. In June 1915 a law had to be passed compelling the military to release 500,000 skilled and semi-skilled men for industry. France suffered from another problem: the large sections of its industrial areas which fell at the outset under German occupation. French production figures reflected these problems. By May 1915 the output of guns was not far in advance of what it had been at the beginning of the

war. By August, however, production of heavy artillery had tripled and by the opening of the Somme campaign production of super-heavy guns showed a five-fold increase over the previous year.

During 1914 and 1915 the weak link in industrial output for the Entente was Russia. Here the disparity between the numbers of men in the army and munitions production was greater than for any other power. But by 1916 Russian industry, contrary to legend, was starting to produce munitions in vast quantities. However, other factors, such as the rudimentary rail network and primitive and corrupt bureaucracy, made it difficult to get munitions to the sector of the front where they were required, or even to get them to the front at all.

If Germany had been fighting France and Russia alone, its dominance in munitions production would hardly have been in doubt. What placed that dominance under threat was the presence of Britain among Germany's opponents. In August 1914, certainly, British intervention did not significantly affect the balance of weaponry. Britain's army (as against its navy) had few munitions suppliers, it possessed no heavy artillery, and it was even short of field guns. Nor was this situation quick to improve. The War Office did set about expanding munitions production, but it soon became clear that its efforts would not be sufficient for the mass army Britain was now raising. So in May 1915 a Ministry of Munitions was formed under the dynamic leadership of Lloyd George, which over a relatively short period transformed the situation. By the time of the Somme battle in mid 1916, Britain had increased its annual output of guns from 90 to 3,200, with the largest percentage of increase occurring in the heaviest types. This was ominous for the Germans, for it presaged the conversion of Britain into a military power of the first order. That, certainly, was hardly the case by July 1916. When the Somme bombardment opened, Britain was still under-equipped in heavy weapons compared to the French and the Germans. In addition, quantity production had been achieved in part by the relaxation of inspection standards, so that many of the shells were defective, either exploding in the guns or littering the battlefield with duds. What was obvious to friend and foe alike was that the British munitions effort would be a mounting factor as the war proceeded.

One other matter was also clear. Britain was using its accumulated wealth and superiority at sea to tap the largest neutral industrial manufacturer in the world. From 1914, orders for raw materials, shells and guns were being placed in the United States. These orders would become, if only in the second half of the war, an important component in Britain's war effort.

VERDUN

The logic of the decisions made towards the end of 1915 meant that Falkenhayn was able to act in February 1916, months before the Allies. After a delay of a week on account of atrocious weather, he launched his blow against Verdun on 21 February. In order to totally devastate the area under attack, he brought his

'THEY SHALL NOT PASS'

The battlefield of Verdun, February–June 1916. The map indicates the dilemma faced by the Germans. If they advanced only on the right bank of the Meuse they could be enfiladed by French artillery on the left bank. If they widened their advance, they dissipated their strength.

stockpile of two million shells and one thousand heavy guns to bear on a front of only 8 miles.

For a moment the German commander appeared to have got his sums right. Despite stubborn resistance by what was left of the French defenders, the Germans in four days advanced a distance of 5 miles on the right bank of the River Meuse. In the process they occupied without a shot the imposing but virtually unmanned Fort Douaumont. German newspapers proclaimed the

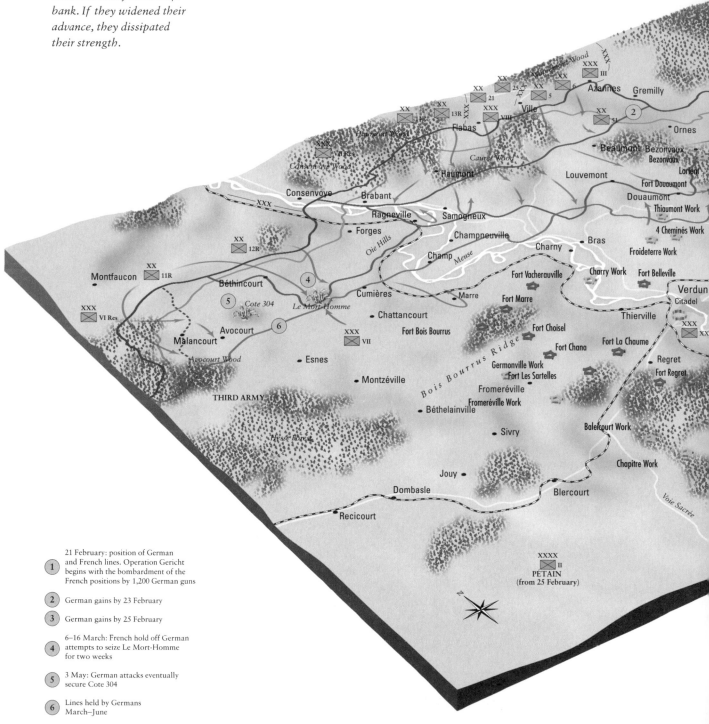

① 21 February: position of German and French lines. Operation Gericht begins with the bombardment of the French positions by 1,200 German guns

② German gains by 23 February

③ German gains by 25 February

④ 6–16 March: French hold off German attempts to seize Le Mort-Homme for two weeks

⑤ 3 May: German attacks eventually secure Cote 304

⑥ Lines held by Germans March–June

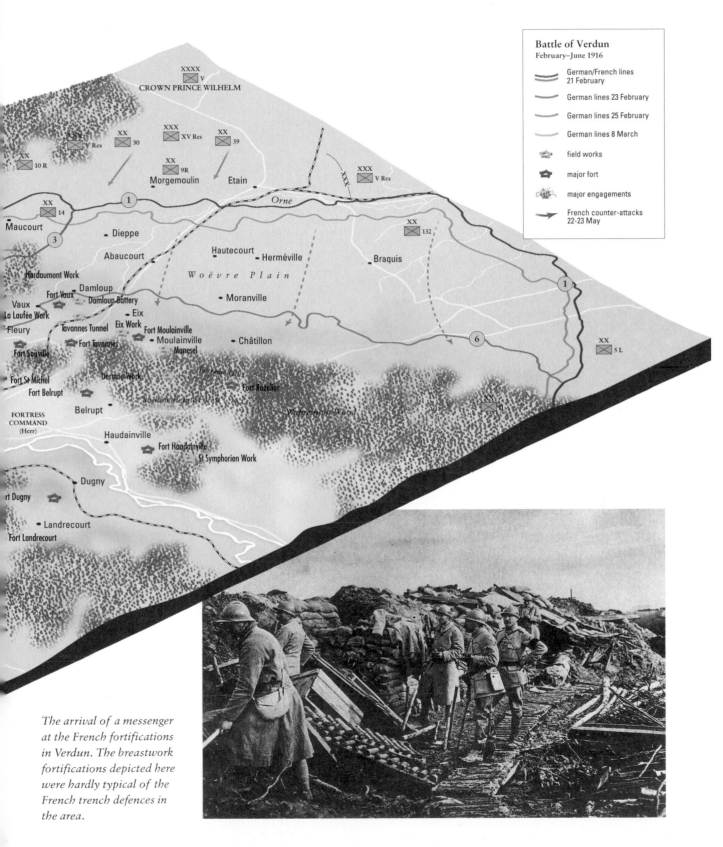

Battle of Verdun
February–June 1916

German/French lines
21 February

German lines 23 February

German lines 25 February

German lines 8 March

field works

major fort

major engagements

French counter-attacks
22-23 May

XXXX
V
CROWN PRINCE WILHELM

XXX
V Res

XX
30

XXX
XV Res

XX
39

XX
10 R

XX
9R
Morgemoulin

Etain

XXX
V Res

Orne

XX
14
Maucourt

3

Dieppe

Abaucourt

Hautecourt

Herméville

Braquis

XX
132

1

Woëvre Plain

Hardaumont Work

Damloup
Fort Vaux
Vaux
Damloup Battery
La Laufée Work

Moranville

Eix
Fleury
Eix Work
Tavannes Tunnel
Fort Moulainville
Fort Tavannes
Moulainville
Manesel

Châtillon

6

XX
5 L

Fort Souville

Fort St Michel
Derume Work
Fort Belrupt

Les Hauts-Champs
Fort Rozelier

XX
fl

FORTRESS
COMMAND
(Herr)

Belrupt

Sartinécourt Forest

Watrenton Wood

Haudainville

Fort Haudainville
St Symphorien Work

Dugny

Fort Dugny

Landrecourt
Fort Landrecourt

*The arrival of a messenger
at the French fortifications
in Verdun. The breastwork
fortifications depicted here
were hardly typical of the
French trench defences in
the area.*

French defeated, and throughout the Fatherland church bells pealed the victory.

It was the last time Falkenhayn's Verdun operation went according to plan. Thereafter, the shortcomings and self-deceptions inherent in his scheme took charge. Two million shells might allow him to make progress on a narrow front on the right bank of the Meuse and to bleed white the French defenders within that limited sector. But they did not constitute sufficient weaponry to bring under attack other crucial areas of the French defensive system. In particular, the area on the left bank of the Meuse, which also happened to lie on the flank of his advancing troops, was left untouched. So, after the capture of Douaumont, the Germans, seeking to get forward, came under increasingly heavy flanking fire from the French artillery massed on the left bank. These guns were being assembled by the master of defensive warfare, Philippe Pétain, who had taken overall command at Verdun. They were now employed to devastating effect. In short order they brought to a halt the German advance. Falkenhayn's momentum, once lost, was never regained.

Hereafter, the Verdun operation became a series of improvisations. In defiance of his original premise, Falkenhayn was obliged to widen his front of assault. While continuing to press forward on the right bank of the Meuse

'The sacred way.' A French motor convoy near Verdun, part of an unending fleet of transports which kept the defending troops supplied despite unrelenting attack.

OPPOSITE: *Philippe Pétain, a novice to battle at the outbreak of war, an army commander by June 1915 and placed in charge of efforts to hold Verdun in February 1916.*

Death in action. During a French counter-attack across the Verdun wastelands, a German cameraman seemingly captures the exact moment when one of the leading infantrymen was struck down by a bullet.

towards his original objectives, he also endeavoured to eliminate the French artillery on the left bank. In the former aspect he made, in the months of March to May, virtually no progress. Indeed, his major accomplishment on the right bank was in fighting off suicidal counter-attacks launched by the newly appointed French sector commander, General Robert Nivelle. On the left bank, the same period witnessed ferocious attrition battles as the Germans strove towards the French artillery concentrations on the Mort-Homme ridge and (extending their front yet again) further west at Cote 304. At last, by the end of May, these positions had been overrun, allowing Falkenhayn to resume operations exclusively on the right bank.

But by this time the offensive had lost all sensible purpose. The blood cost, which according to his plan was supposed to run overwhelmingly to the

disadvantage of the French, was proving almost as great for the Germans. And the projected capture of Verdun – which had it been accomplished by a *coup de main* in February might have meant a great deal – had by now ceased to be a meaningful prize. In the event, anyway, it could not be accomplished. Falkenhayn's forces secured one more small success (the capture of Fort Vaux), and then nothing. The approach of the Allied offensive on the Somme gave Falkenhayn cause – or anyway excuse – for closing down the offensive.

The great German endeavour at Verdun had not been wholly in vain. The French had paid a heavy price, in casualties (about one-third of a million) and in resolve. In ominous incidents, the morale of French forces thrown into the Verdun killing-ground had shown evidence of impairment. But this was insufficient reward from the German point of view. Their losses were almost as great as those of the French. And France now possessed in Great Britain an ally readying itself to take on the main burden of the fighting, while Germany had nowhere to look for help. Moreover, Falkenhayn had promised to resolve both the strategic dilemma facing Germany – east as against west – and also the tactical dilemma: how to drain the manpower resources of the enemy without incurring comparable losses himself. Manifestly, the battle of Verdun had accomplished neither.

The ruins of the city of Verdun. Falkenhayn claimed that the capture of the city was not his objective, but this was a post facto justification for the failure of the Germans to penetrate this far.

BRUSILOV

By mid year, the Central Powers, rather than persisting in attack, were finding themselves obliged to look towards their defences. The concentric operations planned by the Allies at the end of 1915 were now getting under way.

The campaign which the tsar's advisers had agreed to mount on their northern front was slow to develop. The Russian High Command might express eagerness for action against Germany; the commanders who were required to carry out such operations felt otherwise. Their pessimism had been reinforced by an attack they had attempted towards Vilna earlier in the year. Inspired by a moderate success in the area of Lake Naroch in September 1915, further action on this front had been thought to hold some prospect of modest success. Although, on account of the date of its commencement (18 March 1916), it has often been portrayed as an attempt by the Russians to take the pressure off the French at Verdun, this was not its original purpose. Actually the Russians had planned the attack well before the Germans began their assault on the French. Only tardiness in assembling the

General Alexei Brusilov, commander of the Russian forces on the south-eastern front in 1916, studying a map.

weapons and men delayed the operation until mid March. By then it was too late. The thaw reduced Russian movements to a crawl and laid them open to compelling counter-attack. This abortive endeavour put paid to any remaining enthusiasm in Russia's northern command for action against the Germans. They concluded that they should not try again until supplies of heavy ammunition had reached formidable levels.

What caused Russia's summer offensive ever to get under way was action by the recently appointed commander on the southern (Austro-Hungarian) front, General Alexei Brusilov. He argued against his forces playing only a minor part in forthcoming operations and proposed to act simultaneously with the offensive in the north and on an extensive scale. To generate interest in this proposal among the tardy northern commanders, he offered to proceed without any great increase in his resources. Spurred, if with continued hesitation, by this apparently sacrificial proposal, the northern command agreed that Russia's campaign against the Central Powers would proceed in line with the Chantilly agreements.

As it happened, Brusilov acted even in advance of his colleagues. This was

because of events on the Italian front, where Cadorna was finding himself in need of help. Back in March the Italian commander had moved against the Austro-Hungarians on the Isonzo for the fifth time, with the same discouraging results as in 1915. His plans for a further attempt in the summer were pre-empted by the Austro-Hungarians, who struck against the Italian left flank in the Trentino. Conrad's plan was to drive as far as Venice, cut off the bulk of the Italian army on the Isonzo and force Italy out of the war.

Codenamed the Punitive Expedition and employing massive artillery support, the Austro-Hungarian offensive was launched on 15 May 1916. Conrad enjoyed considerable early success. Two lines of defences fell to his forces, 400,000 Italians were taken prisoner, and for a moment it seemed that he might break out into the Venetian plain. It was this event that led the Italians to appeal to Brusilov to mount diversionary operations.

In the event, as the advance of the Austro-Hungarian forces carried them beyond their artillery support, Conrad's momentum slackened anyway. But by then General Brusilov had moved and was making a decided impact on the south-eastern front against the Habsburg troops.

Brusilov is sometimes rated the outstanding commander of the First World War. This appellation, which others would ascribe (improbably) to Lettow-Vorbeck in East Africa or Allenby in Palestine or even Lawrence in Arabia, ignores the crucial point. Brusilov, like these others, was not fighting on the main front or against the main adversary. Certainly, his manner of attacking defied the accepted method of conducting an offensive in this war. Despite the limitations of manpower and weaponry imposed on him, he did not concentrate his forces against a particular sector but attacked along the whole front. Combating Habsburg forces seriously denuded for the Trentino operation, and whose morale anyway was far from secure, this yielded astonishing early success. Commencing on 4 June, his operation in a month advanced 60 miles along his whole front and took 350,000 prisoners. But to identify this as a more inspired and imaginative way of conducting war is to miss the point. Employed against the Germans, his methods would have swiftly brought his offensive to grief.

As it was, Brusilov's rate of advance could not be maintained. Initially his opponents were misled by the fact that the bulk of Russia's forces were elsewhere, and by the width of front along which he chose to attack. So they were taken by surprise and their fragility exposed. But these same circumstances denied Brusilov the reserves of men and weaponry, and the means of transport, needed to exploit success. Further, victory over the Austro-Hungarians was eventually bound to be

Russian troops near Lake Naroch. These forces were part of an abortive offensive on the northern section of the Eastern Front in March 1916.

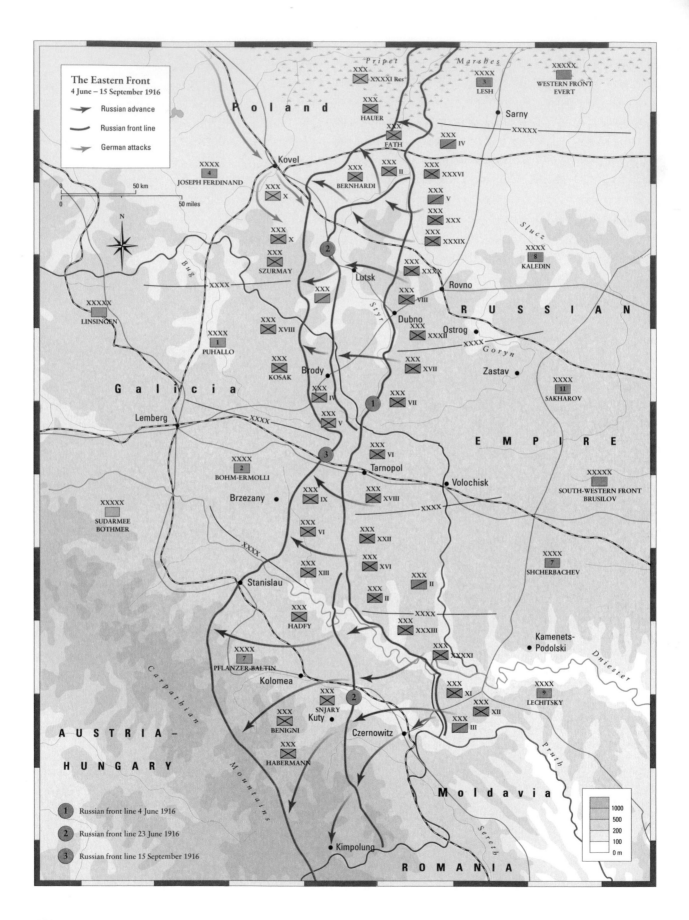

The Eastern Front
4 June – 15 September 1916

→ Russian advance

↷ Russian front line

→ German attacks

0   50 km
0   50 miles

N

Pripet   *Marshes*

P o l a n d

XXXXX
WESTERN FRONT
EVERT

XXX XXXXI Res

XXXX
3
LESH

XXX
HAUER

Sarny

XXX
FATH

XXX
IV

XXXXX

XXXX
4
JOSEPH FERDINAND

Kovel

XXX
BERNHARDI

XXX
II

XXX
XXXVI

XXX
X

XXX
5
V

XXXX
8
KALEDIN

XXX
X

XXX
XXXIX

Slucz

XXX
SZURMAY

2

XXX
XXX

Bug

Lutsk

XXX
VIII

R U S S I A N

XXXX

XXX

Styr

Rovno

XXXXX
LINSINGEN

XXX
XVIII

Dubno
XXX
XXXII

Ostrog

Goryn

XXXX
1
PUHALLO

XXXX

XXX
KOSAK

Brody

XXX
XVII

Zastav

XXXX
11
SAKHAROV

G a l i c i a

XXX
IV

XXX
VII

1

XXX
V

Lemberg

XXXX

E M P I R E

XXXX
2
BOHM-ERMOLLI

3

XXX
VI

Tarnopol

XXXXX
SOUTH-WESTERN FRONT
BRUSILOV

Brzezany

XXX
IX

XXX
XVIII

Volochisk

XXXXX
SUDARMEE
BOTHMER

XXX
VI

XXX
XXII

XXXX

XXX
XIII

XXX
XVI

XXXX
7
SHCHERBACHEV

Stanislau

XXXX

XXX
II

XXX
II

XXX
HADFY

XXX
XXXIII

Kamenets-
Podolski

Dniester

XXX
XXXXI

XXXX
7
PFLANZER-BALTIN

Kolomea

XXX
XI

XXXX
9
LECHITSKY

XXX
SNJARY

XXX
XII

XXX
BENIGNI

Kuty

2

XXX
III

Pruth

XXX
HABERMANN

Czernowitz

A U S T R I A –

M o l d a v i a

H U N G A R Y

Sereth

Carpathian

Mountains

1000
500
200
100
0 m

1   Russian front line 4 June 1916

Kimpolung

2   Russian front line 23 June 1916

3   Russian front line 15 September 1916

R O M A N I A

arrested by Germany's determination to shore up its ally. As increasing numbers of German forces were moved to the southern sector, the conditions underpinning Brusilov's success departed.

The Russian High Command, for its part, was uncertain how to respond to Brusilov's early triumph. Should it divert forces to his front, or should it set in train its tardy northern offensive promised against the Germans? Late in July, the decision was taken to proceed with the latter. The outcome was no happier than on previous occasions. In no time Russia's offensive against the Germans on their

THE BRUSILOV OFFENSIVE 1916

This map illustrates the enormous length of front over which the Russians attacked. As a result of the offensive Brusilov was hailed as a major commander. All his success proved, however, was that a victory against the Austrians was relatively easy to attain. When the Germans entered the equation, Brusilov was no more successful than other commanders in 1916.

BELOW: *Russian reserves 1916: some of the vast manpower preparing for Brusilov's attack, an episode that briefly promised to be a decisive campaign on the Eastern Front.*

'The opportunity for which Romania had so long watched had not only come. It had gone.' (Winston Churchill) Following a battle near Focsani, Romanian prisoners are marched into captivity. German forces would overrun Romania in nine weeks.

northern front, which had been regarded as the tsar's main contribution to Allied strategy, had been stopped in its tracks.

Brusilov's days of glory were also about to end. The rulers of Romania, aroused by the apparently imminent demise of the Habsburg empire, chose this moment to enter the fray on the Allied side. The event was wildly welcomed by the Entente Powers and struck dismay among their opponents. The kaiser declared the war lost, and at last consented to the dismissal of Falkenhayn and his replacement by the overbearing easterners, Hindenburg and Ludendorff. Yet nothing about Romania's circumstances warranted such extreme reactions. The Romanian regime possessed none of the resources, motivation or competent

military advisers required in a struggle of this magnitude. Its defeat at the hands of the Germans was swift. And Brusilov – who by now had problems enough with the German forces sent to aid the Habsburgs – was obliged to involve himself in Romania's operations and so get caught up in its defeat.

Before the year had ended, all the territory which Brusilov had captured was back in the hands of his enemies. His huge casualties – some 1.4 million – had left his army desperately depleted and disheartened. He had laid bare the fragility of Austria–Hungary, but also Russia's inability to take lasting advantage of it. Above all, Russia's summer offensive had driven home the message that, with the resources available to it and under the existing structures of government and command, the tsar's regime could not overcome the challenge of Germany.

THE SOMME

The great question confronting the Anglo-French offensive on the Somme in the high summer of 1916 was whether it could accomplish a different result from the offensives so far conducted that year. The campaign had been conceived as a predominantly French affair. The British were intended to contribute either by wearing down the Germans in advance of the main attack (Joffre's idea) or by

Trench warfare on the Western Front. A French sniper using a periscope rifle fires on the enemy while keeping his head below the parapet.

acting alongside the French as the lesser component of a single concerted operation (Haig's choice). Haig's determination ensured that it would not be the former. Falkenhayn's action at Verdun determined that it would be neither.

As the killing at Verdun proceeded, it became evident that the French contribution to the Somme must be much diminished. Haig's army, not Joffre's, would now be the major participant. Nevertheless, the original conception remained. The offensive was to be preceded by a bombardment so devastating that it would overwhelm the formidable defences built by the Germans on the Somme during the previous two years. The attack would be delivered on a wide front (20 miles), so that in the centre the assaulting forces would not be subjected to the flanking fire which had so unhinged German designs at Verdun. And Haig was not just contemplating the overrunning of several lines of German trench. Rather he aimed to rupture the entire German defensive system, pour the cavalry through the breach into open country, and reinstate a war of movement and strategic accomplishment.

The notion that the attack must be on a wide front to avoid the injury of flanking fire was soundly based. But it carried two important corollaries which the British Command was not engaging. One was the formidable quantities of ammunition required to demolish the extensive German defences confronting the attacking forces. The other was the degree of skill required of Haig's artillerymen to deliver the bombardment with the required effect. Haig and his associates spoke as though they had taken these matters on board. But this was delusion. Compared with the munitions available for previous British offensives, the

The campaign on the Somme: a panoramic view of Fricourt valley after its capture by the British early in the battle.

quantities of guns and shells at Haig's disposal may have been large. Compared with the size of the operation now being undertaken, they were not.

It was not just in numbers that the Somme bombardment was deficient. There were serious inadequacies in quality, type and manner of delivery. As already mentioned, the Ministry of Munitions had made a number of questionable decisions. To increase output, the Ministry had abandoned quality control, with the consequence that many shells failed to explode or detonated in their own guns. Further, an inappropriate proportion of the shells consisted of shrapnel, which had a role to play in cutting the enemy wire or dealing with enemy forces in the open, but was ineffective against troops in deep dug-outs. Even most of the high explosive shells being employed lacked the power to penetrate into these dug-outs.

A battery of 8-in. howitzers in operation near Mametz, August 1916. From August, British artillery resources far exceeded those employed by Germany on the Somme.

There could be but one consequence. The British bombardment that preceded the Somme offensive was inadequate to knock out key elements in the German defence, particularly the machine-guns in close proximity to the front and the mass of heavy guns well behind it. In short, an attack on the scale that Haig was contemplating was doomed in advance.

On 1 July 1916 the notion that the Somme campaign would differ from all

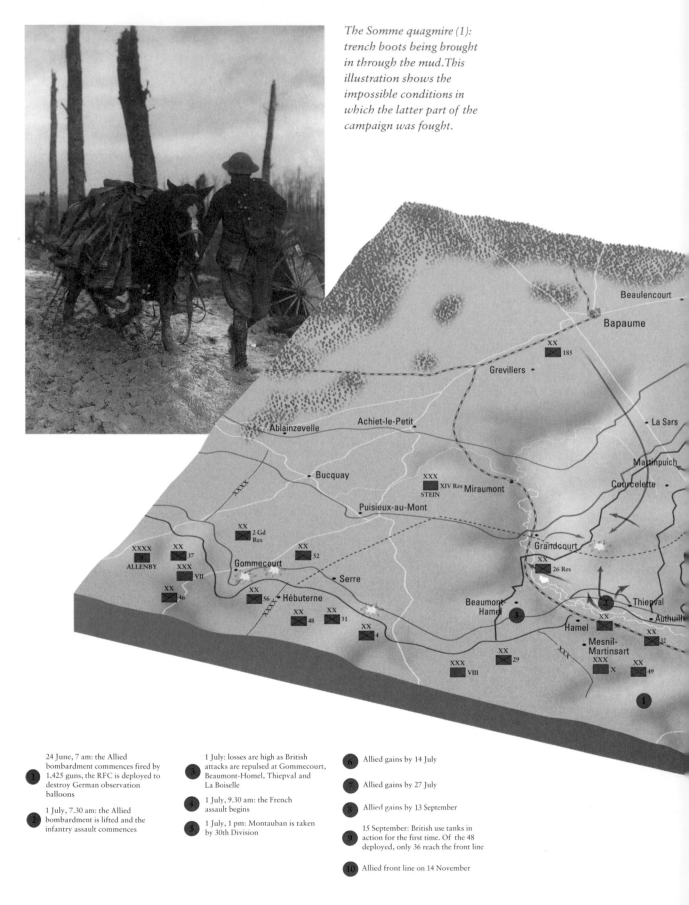

*The Somme quagmire (1):
trench boots being brought
in through the mud. This
illustration shows the
impossible conditions in
which the latter part of the
campaign was fought.*

Beaulencourt

Bapaume

Grevillers

XX 185

La Sars

Achiet-le-Petit

Ablainzevelle

Martinpuich

Courcelette

Bucquay

XXX
XIV Res Miraumont
STEIN

Puisieux-au-Mont

XXXX

XX
2 Gd
Res

Grandcourt

XX
26 Res

XXXX
3
ALLENBY

XX 37

XX 52

Gommecourt

XXX
VII

Serre

Thiepval

XX 2

XX
46

XX 56

Hébuterne

Beaumont-
Hamel

XX 3

Authuill

Hamel

XX 36

XXXX

XX 48

XX 31

Mesnil-
Martinsart

XX 32

XX 4

XXX

XX 29

XXX
VIII

XXX
X

XX 49

XX
1

24 June, 7 am: the Allied
bombardment commences fired by
1,425 guns, the RFC is deployed to
destroy German observation
balloons

1

1 July, 7.30 am: the Allied
bombardment is lifted and the
infantry assault commences

2

1 July: losses are high as British
attacks are repulsed at Gommecourt,
Beaumont-Homel, Thiepval and
La Boiselle

3

1 July, 9.30 am: the French
assault begins

4

1 July, 1 pm: Montauban is taken
by 30th Division

5

Allied gains by 14 July

6

Allied gains by 27 July

7

Allied gains by 13 September

8

15 September: British use tanks in
action for the first time. Of the 48
deployed, only 36 reach the front line

9

Allied front line on 14 November

10

other Western Front offensives, and would witness the restoration of a war of movement, died at a single blow. Haig's 400 heavy and 1,000 field guns proved quite insufficient to crush a defence of 4,500 yards in depth. So when the British infantry moved forward to cross no-man's-land they were met with a hail of machine-gun bullets and high explosive shells and in some places belts of uncut barbed wire. In the course of the day Haig's army sustained 57,000 casualties, including a chilling 20,000 dead. Their reward was negligible: an advance on the right of the attack as far as the German front line and a small distance beyond,

THE SOMME CAMPAIGN

The most costly battle ever fought by the British army. Note that Bapaume, an objective for the first day, remained in German hands at the end of the battle.

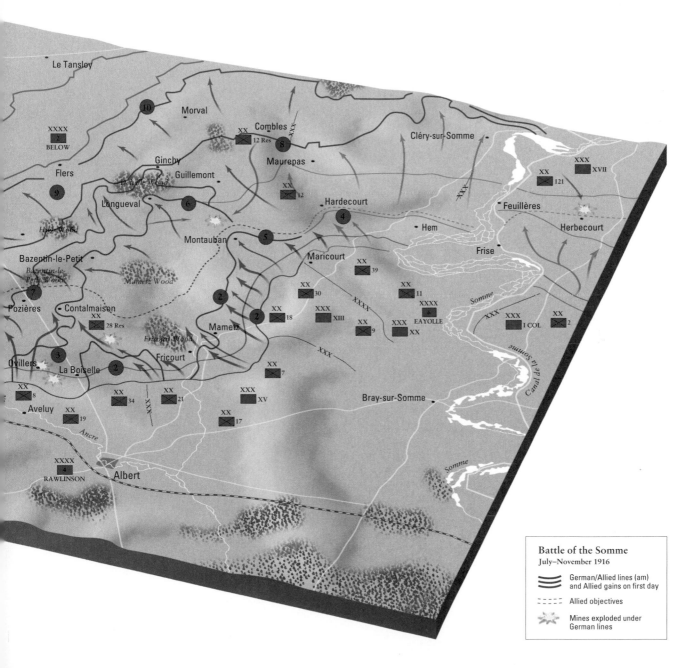

Battle of the Somme
July–November 1916

German/Allied lines (am) and Allied gains on first day

Allied objectives

Mines exploded under German lines

and a complete absence of advance both on the left and in the key area in the centre. To the south, the French, using many more guns per mile of front attacked, did rather better. They overran the entire German front system and advanced some way towards the second.

There could be no question that this dismal opening to the Somme campaign would also be its termination. The French were still under pressure at Verdun; the British had prepared for this moment for two years. The great endeavour could hardly be abandoned.

But if the continuation of the campaign was not a matter for reconsideration, its hoped-for purpose might be. The events of 1 July seemed to call in question the whole notion of a breakthrough. If the power of defensive weaponry was ever to be subdued, it could only occur within the range of high explosive shells fired with greater skill and in far larger numbers than had so far been possible.

What happened on the Somme battlefields between 2 July and mid November, when the onset of winter tardily brought the campaign to a close, hardly showed that the Allied Command was taking these considerations into account. There were long spells, certainly, when it looked as though Haig was revising his lofty aspirations. Yet on closer examination this proves not to be the case. Rather, although there were periods which witnessed small-scale, piecemeal operations directed towards limited objectives, these were always intended as preliminaries to a projected larger purpose. That is, a succession of small endeavours was meant in time to produce the

British Mark 1 Tank, of the type which first saw battle on the Somme on 15 September 1916. Tanks would prove of modest utility to the Allied side for the remainder of the war.

conditions whereby the infantry and the artillery would overwhelm the enemy's defences and the cavalry would sweep forward.

The campaign, in the aftermath of Day One, fell into three phases. The first occupied July and August. Offensive operations continued, but apparently with little plan or purpose. There was one well-managed advance on a limited front, employing an overwhelming bombardment, on 14 July. But the cavalry action that was supposed to follow from it was a fiasco from the start. For the rest, these weeks witnessed only narrow-front, uncoordinated operations which attracted maximum enemy firepower. As a consequence, between 2 July and the end of August Haig's army suffered greater casualties than on 1 July and did not capture much more ground.

Certainly there was another side to all this, as had scarcely been the case on

Ammunition limber (i.e. carrier or cart) passing through the ruined village of Longueval, September 1916.

1 July. German regimental accounts of the battle, along with the German Official History, make it clear that for the kaiser's soldiers the Somme campaign was one of the worst ordeals of the war. In particular, German forces were appalled by the sheer volume of artillery fire that the British could now bring to bear on them. So between 2 July and mid September British gunners fired over 7 million shells into the German positions on the Somme. This was far in excess of what the outnumbered German artillerymen could fire back. Not surprisingly, the battle became known as the '*materialschlact*' – the munitions offensive.

It is also worth noting that, in part, the ordeal suffered by the German troops was inflicted by their own command. During the middle period of the battle, it was German policy to recapture by counter-attack every yard of ground gained by the British – whether it was of tactical value or not. So in some measure the British policy of uncoordinated attacks was matched by an equally questionable German counter-attack policy. By these means the Germans managed, at least to some extent, to equalize the casualty lists.

By September Haig was prepared to make his next large effort. A new weapon and new methods of employing established weapons were at hand. As regards the former, at last he had available a quantity of the eagerly awaited tanks. Although slow-moving, mechanically unreliable and vulnerable to shelling, they were at least fairly immune from rifle and machine-gun fire and so enjoyed a chance of getting across no-man's-land. But far more significant, if less spectacular (and so less likely to enter into the war's mythology), was the new method of employing artillery. British gunners were now capable of firing a creeping barrage of shells in coordination with the forward movement of their infantry. This went some way towards neutralizing trench defenders, but not enemy artillery, during the perilous period of advance across no-man's-land.

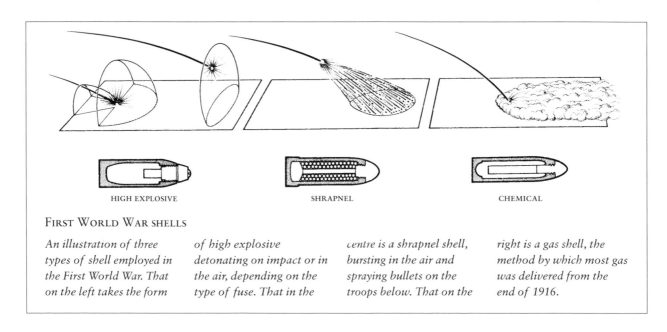

HIGH EXPLOSIVE   SHRAPNEL   CHEMICAL

FIRST WORLD WAR SHELLS

An illustration of three types of shell employed in the First World War. That on the left takes the form *of high explosive detonating on impact or in the air, depending on the type of fuse. That in the* *centre is a shrapnel shell, bursting in the air and spraying bullets on the troops below. That on the* *right is a gas shell, the method by which most gas was delivered from the end of 1916.*

The unrelenting war of the guns: a heavy howitzer battery.

Regrettably for Haig's forces, on the first occasion when both tanks and creeping barrage were used in the attack, they rather cancelled each other out. Haig planned 15 September as his second attempt at a great breakthrough. The chosen front was wide, and the cavalry held at the ready. But the means of proceeding proved faulty. In order to protect his new weapon, the tank, from his own artillery shells, the creeping barrage was not fired ahead of his infantry in the areas – which happened to be the most formidable centres of resistance – where the tanks were being employed. This was a misjudgement. The tanks, despite spasmodic successes, proved too untrustworthy mechanically and too vulnerable to shellfire to get the infantry forward. Added to this, Haig's yearning for action by the cavalry caused him to spread his guns too thinly over too wide a front.

In the outcome, therefore, the action of 15 September produced some piecemeal advances but no redemption for the failure of 1 July. Ten days later, without the use of tanks, which were mostly incapacitated, Haig did slightly

OVERLEAF: *British troops in the final phase of the Somme campaign: the scene on the morning of the battle of Morval, 18 October 1916, during a rare spell of fine weather.*

better. An entire section of the German front defences was captured by dint of firing a continuous creeping barrage and by strictly limiting the object of the attack. But this success signified nothing major. By this time, the Germans had constructed a further series of defences which stretched for miles in front of the Allied armies.

It was now late in the year. Rain was falling and would soon transform the battlefield into a quagmire. Yet Haig persisted in the Somme campaign – either out of habit or from some conviction that the Germans were at the end of their tether. By mid November it was his own forces that were showing alarming signs of depletion in numbers for which there were insufficient immediate replacements. So in the midst of rain and bog he was at last forced to call a halt. The French, whose role had been limited to keeping pace with the British right flank, also ceased operations.

As far as can be gauged, Allied casualties for the campaign numbered 650,000, as against 400,000 German. This could only be considered to the advantage of the Allies if it were calculated that their resources of manpower and industrial capacity greatly exceeded those of their opponents. Such a calculation, given the uncertain and varying combinations of forces arrayed against each other by the end of 1916, constituted the merest guesswork.

The Somme quagmire (2): horses up to their knees in mud bringing up supplies of ammunition, November 1916. Conditions like this finally persuaded Haig to call off the campaign.

END OF THE DAY

All told, the battles of 1916 had brought much grief and small rewards to both sets of combatants. Germany had once more held off a challenge from Russia, and had acquired the abundant and badly needed raw materials of Romania. Its major ally, although surviving against Italy, would clearly not have outlived assault by Russia but for German intervention. And Germany had failed first in its attempt to defeat France (on land) at Verdun, and then Britain (at sea) in the battle of Jutland on 31 May. Moreover, it had endured heavy losses in the Anglo-French campaign on the Somme.

For the Allies, the year offered no great reassurance. Russia, despite a happy interlude on its less important front, had ended the year as severely rebuffed as it had opened it, and with the question seriously looming as to whether – anyway with its existing regime – it could remain in the war at all.

France, in defence and then in offence, had once again endured a scale of casualties which plainly could not be continued indefinitely, and had certainly not advanced significantly towards victory. Britain's situation was somewhat happier; or, more accurately, was not yet this grievous. Britain retained command of the sea and access to a wide range of resources. It was approaching, by year's end, a high state of mobilization. It had not yet reached the drained condition of allies and adversaries. Yet it had still failed to divine the means by which the conflict could be brought to a successful conclusion. And its first major involvement in continental-scale fighting had brought it heavy human losses and few rewards. As the historian G. H. Gretton wrote in 1930, 'No one who lived through that time will ever forget the casualty lists of the Somme.'

CHAPTER FIVE

1917

Canadians at Vimy Ridge turn a German gun against its former owners, April 1917. The capture of this commanding position was one of the major feats of the war. However, no further progress was made against the Germans in this sector until the last weeks of 1918.

1917

CHOICES

In determining their strategy for 1917, Germany's leaders engaged in a major reappraisal. The contrasting experiences of the kaiser's forces on the Eastern and Western Fronts in 1916, and the supersession of Falkenhayn by Hindenburg and Ludendorff, ensured that for Germany the year 1917 would not be *la même chose*.

The new command soon came to the conclusion that there must be no more wearing-out offensives in the manner of Verdun. They therefore decided to go over to the defensive in the west, even to the extent that they would surrender territory in order to occupy a shorter, stronger defensive line – the Hindenburg position. This voluntary abandonment of territory which the German army had fought so devotedly to retain during the battle of the Somme was carried out between February and April.

In the east Hindenburg and Ludendorff contemplated no great action. Russia had been so severely dealt with in 1916 that it represented no immediate danger.

Aftermath of the German retreat, March 1917: British troops at last occupy former enemy positions at Serre which had been an objective for the leading troops on 1 July 1916.

But the new commanders did have two large purposes in mind. They would carry out a programme of extensive industrial mobilization, harnessing the entire German economy for the production of war matériel. And they would attempt to bring Britain to its knees by a campaign waged not on land but at sea, and directed not against the British fleet but against merchant shipping – neutral as well as Allied. The U-boat would be employed to starve Britain into submission.

Germany's plans for 1917 certainly possessed some elements of novelty. That could hardly be claimed for those of the Entente. In the east, despite the growing decrepitude of the tsar's forces, offensives were contemplated in the north and in the south. But this time the main attack would be directed against Austria–Hungary. The commander would be Brusilov.

On the Western Front, notwithstanding the disappointments of 1916, another series of offensives was also in prospect. In France there was now a new commander. General Nivelle, who late in 1916 had successfully carried out a number of limited counter-offensives at Verdun, had replaced Joffre as French commander-in-chief. The conviction arose that here was a new man holding the secret of a wide-ranging success.

In Britain too there had been change, but at the political rather than the military level. Late in 1916 the long-standing Liberal Prime Minister, H. H. Asquith, had been replaced by David Lloyd George leading a coalition government which, although including Liberal and Labour elements, was made up largely of Conservatives. The attitude of the new Prime Minister to new offensives was ambiguous. He was devoted to the unrelenting prosecution of the war – 'a fight to a finish, to a knock out'. Yet he was no devotee of Haig, whom he regarded as stubborn, unimaginative and cruelly wasteful of his soldiers' lives. He was determined that the new year would see no sacrifice of British troops on the scale of 1916.

To reconcile these seemingly contradictory attitudes, Lloyd George came up with a plan to employ Italy in early 1917 to do the lion's share of the fighting for the Allies. The plan foundered on Italian reluctance to adopt this sacrificial role. Lloyd George then turned to Nivelle and his proposal for a great French war-winning offensive, on the questionable ground that French generals could succeed where British generals could not. With immoderate

General Robert Nivelle, who came to prominence in the course of the Verdun campaign and was appointed French commander-in-chief in late 1916. His 'method' of accomplishing the rupture of the German lines on the Chemin des Dames in 1917 made only a small dent in their position and reduced his own army to mutiny.

The British contribution to the Nivelle offensive: an 18-pounder battery in action in the open during the battle of Arras, April 1917.

enthusiasm, the British Prime Minister adopted Nivelle's proposal for a major French offensive in the spring delivered on the Chemin des Dames. It was to be aided further north by a supporting British offensive out of Arras. More remarkably, he agreed that the overall commander henceforth of British forces on the Western Front would be Nivelle and not Haig. This proposal, later modified (under protest) so that Nivelle would control British forces only for the duration of this one offensive, poisoned relations between Lloyd George and British headquarters for the remainder of the war. Nevertheless, Haig was obliged to acquiesce. Thus was the stage set for the first phase of momentous, not to say tragic, events on the Western Front in 1917.

THE U-BOAT CAMPAIGN

The first of these varied offensives devised by the opposing sides got under way in February. Germany unleashed its campaign of unrestricted U-boat warfare against Allied and neutral merchant shipping. The course of the war at sea is related in another volume in this series and will not be dealt with in detail here. Suffice it to say that the U-boat campaign produced spectacular early results, yet never came within sight of achieving its objective. It was founded on a calculation concerning the number of tons of shipping that would have to be sunk to strangle

Britain which was never anything more than a wild guess. And anyway, the toll of heavy loss among merchant ships imposed by the U-boats in the first half of 1917 could not be maintained. British counter-measures, especially the adoption of convoys, soon deprived the German submarines of their easy prey.

The only enduring consequence of this campaign was entirely to Germany's disadvantage. President Woodrow Wilson, as he had made clear would be the case, treated the destruction of US merchant ships as a *casus belli* and entered the war on the Allied side. This brought an appreciable accretion of naval power to the Entente, and also removed any danger that American credit to the Allies might soon come to an end. What American entry into the war did not bring to the Allies was any immediate infusion of fresh troops. When it entered the war in April 1917 the USA possessed but a small army, and thereafter it made only deliberate pace in mobilizing its vast manpower resources. Moreover, its commander, General John Pershing, was determined to keep his forces out of battle until they could be employed as an independent unit.

The consequences of unlimited U-boat warfare: American President Woodrow Wilson brings the USA into the war against Germany, April 1917.

Exit Russia

If Germany's major offensive for 1917 made an early impact and then languished, Allied operations that year did not enjoy even that much good fortune. They simply languished.

The proposed Russian campaign for the first half of the year never reached the starting line. Promised for 1 May, the Russian Command by early March had accumulated sixty-two divisions and much firepower. But already by February it was evident that the Russian army's capacity for offensive action was crumbling. Transportation was approaching collapse, morale was falling, and revolutionary sentiments were making themselves felt. During March and April, desertions reached two million.

Even more decisive were events on the civilian front. The tsar's regime had become so despised that many even among its supporters recognized that a change must be made. The final impetus came from the capital, Petrograd. Widespread strike action broke out, occasioned largely by food shortages, and the armed forces refused to suppress it. The tsar was forced to abdicate and a

*Russia's last throw:
Alexander Kerensky, having
emerged as Russian premier
following the collapse of the
tsarist regime, mounts what
will prove to be the last
Russian offensive action of
the war.*

constitutional government, liberal in complexion, was formed to fill the void. But its hold on power was precarious, and in the large cities it was challenged by the formation of workers' councils. There seemed to be a compelling case for suspending offensive military action until the new regime had consolidated its position. But this was not the view taken by the leader of the Provisional Government, Alexander Kerensky.

Kerensky held that only a major victory in the field could establish the credibility of his administration, both at home and with his Anglo-French allies. Brusilov was instructed to improvise an offensive against the Austro-Hungarians.

The attack opened on 18 June. Spearheaded by specially selected shock troops, it prospered briefly. Then the Germans, as ever, intervened, the shock troops were eliminated, and Russian follow-up forces revealed no great enthusiasm for the endeavour. Within three weeks the Kerensky offensive was at an end and the Russian army in headlong retreat.

The Russian collapse caused the German Command to revise its views about action on the Eastern Front. At the beginning of September, on their northern sector, German forces employing new offensive tactics (soon to be used against the Italians and the British) struck at the Russian right flank. In no time they had accomplished the capture of the major port of Riga on the Baltic.

At this point the morale of Russia's armed forces disintegrated entirely. What remained of authority in the upper echelons – where Brusilov had been sacked and General L. G. Kornilov taken his place – was directed not against the enemy but against Russia's civilian government. Confronted with these acts of disloyalty, with the challenge of revolutionary forces in the cities, with peasant demands for a redistribution of land, and with the break-up of the army, the liberal government of Kerensky collapsed.

The disintegration of the Russian army: Russian forces in retreat in Galicia, September 1917.

The USA enters the war. Soldiers of the New York National Guard in a training camp. A year later, as members of the 27th US Division, these men would participate alongside the Australians in attacking the Hindenburg Line.

Power passed, at least in the cities, to the revolutionary Bolsheviks, who believed that the moment was at hand to generate a social upheaval throughout Europe and to bring the war to a halt from below. As a first stage they convinced themselves that abandonment of the war by Russia would cause rank and file German soldiers to defy their commanders and down arms. These expectations were to be savagely disappointed.

NIVELLE'S DAY

As the Western Allies watched helplessly the military collapse and political dislocation of Russia, their own experiences were providing no comfort. Given the strength and coherence of their political and social structures, neither France nor Britain was brought close to disintegration by their misfortunes. But a year of stalled enterprises and telling setbacks, if offset by survival at sea and the eventual promise of American participation, gave rise to serious doubts as to when and how the war could be won.

France's great endeavour of 1917, the Nivelle offensive on the Chemin des Dames in April, proved a disaster. Only the British contribution to this campaign, intended to be its lesser component, yielded any reward, and that only in its dramatic opening stages. On 9 April Canadian forces assailing Vimy Ridge and

*German troops cross the
River Dvina, 2 September
1917, on the way to
capturing the Russian port
of Riga.*

British forces attacking out of Arras captured a considerable vantage point and achieved a greater advance in a single day than any previous British operation on the Western Front. It so happened that the attacks were delivered against positions where the enemy – placed on forward slopes and with their reserves held too far back – were poorly situated. But it was also the case that the mass of artillery now available to Haig's forces, and the manner of its employment, gave proof that the industrial resources of Britain and its Dominions were now, finally, being fully geared towards military purposes, and that the skill required to make effective use of the resulting weaponry was also being properly developed and utilized.

What the events of 9 April did not reveal was that any means existed to convert initial advances, however noteworthy, into breakthrough. The cavalry proved as ill-equipped to exploit success on this occasion as on any other. And the

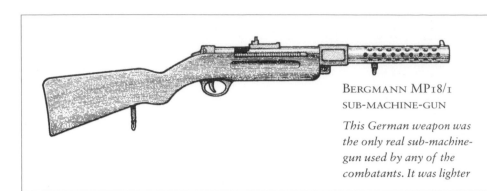

BERGMANN MP18/1
SUB-MACHINE-GUN

This German weapon was the only real sub-machine-gun used by any of the combatants. It was lighter than the British Lewis gun, easier to load, and fired a 32-round magazine at a rate of 540 rpm. It was, however, only introduced in 1918 and in limited numbers. 'MP' stands for 'machine pistol', an indication of the gun's portability.

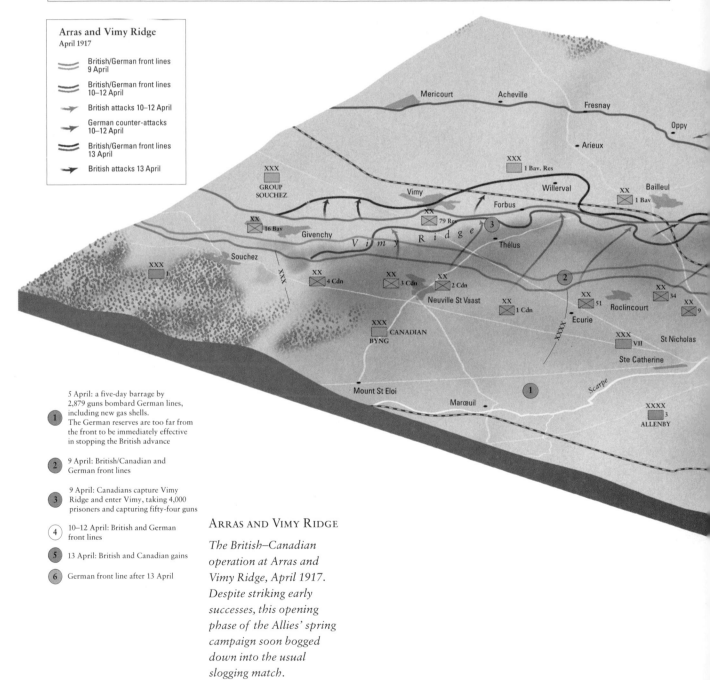

Arras and Vimy Ridge
April 1917

British/German front lines
9 April

British/German front lines
10–12 April

British attacks 10–12 April

German counter-attacks
10–12 April

British/German front lines
13 April

British attacks 13 April

1. 5 April: a five-day barrage by 2,879 guns bombard German lines, including new gas shells. The German reserves are too far from the front to be immediately effective in stopping the British advance

2. 9 April: British/Canadian and German front lines

3. 9 April: Canadians capture Vimy Ridge and enter Vimy, taking 4,000 prisoners and capturing fifty-four guns

4. 10–12 April: British and German front lines

5. 13 April: British and Canadian gains

6. German front line after 13 April

ARRAS AND VIMY RIDGE

The British–Canadian operation at Arras and Vimy Ridge, April 1917. Despite striking early successes, this opening phase of the Allies' spring campaign soon bogged down into the usual slogging match.

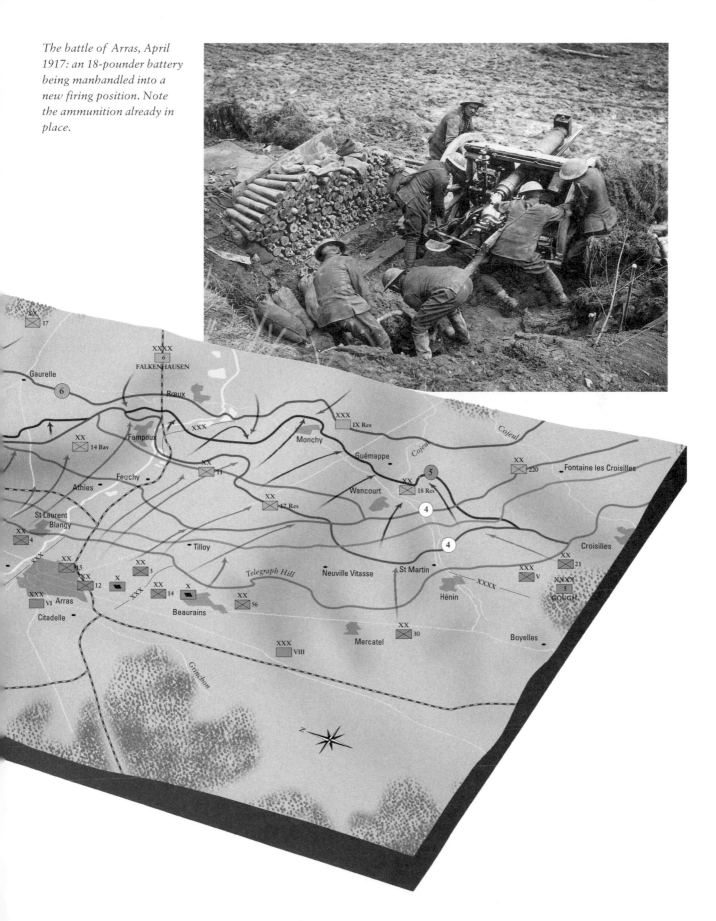

The battle of Arras, April 1917: an 18-pounder battery being manhandled into a new firing position. Note the ammunition already in place.

Behind the lines at the Chemin des Dames. French positions seen here on a reverse slope, out of direct line of fire of the German guns.

artillery, having contributed so largely towards early advance, simply could not be moved forward fast enough, or registered on fresh targets competently enough, to enhance this success. Within days Haig's forces at Arras were simply hammering away, at mounting cost to themselves and with diminishing reward, in the too-familiar manner of the year before.

The British offensive was followed on 16 April by the opening of Nivelle's attempt to seize the Chemin des Dames. But here early success of any note eluded the French. Nivelle had promised a complete rupture in the enemy front within 48 hours, or the offensive would be halted. Yet nothing about his methods offered early breakthrough. His artillery innovations certainly enhanced offensive techniques in a way similar to those now being used by the British. But on the Chemin des Dames these were largely countered by a German withdrawal to reverse slopes and the employment of a well-manned system of defence in depth. As a consequence, although Nivelle did manage to get forward in places ('more slowly,' as he

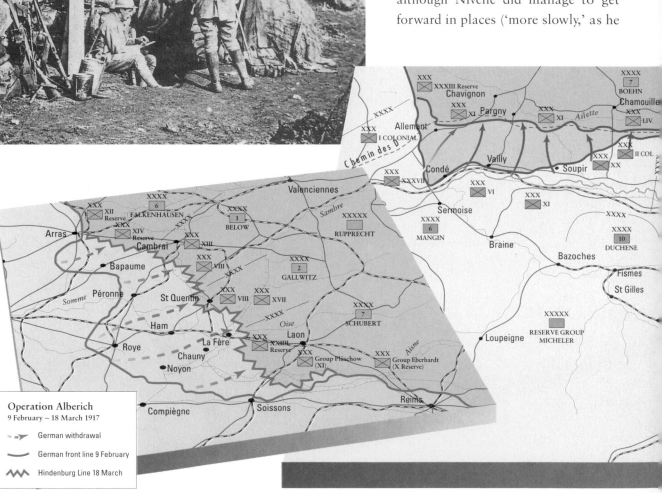

Operation Alberich
9 February – 18 March 1917

➤ German withdrawal

⌒ German front line 9 February

ᗘᗘᗘ Hindenburg Line 18 March

himself admitted, 'than we had hoped'), what was most evident was that this process amounted to no more than a repetition of Joffre's 'nibbling' operations of the past.

Had this been 1915 or even 1916, such an appearance might not have signalled calamity for Nivelle. But the drain on French manpower had now reached a critical point. Already, episodes during the Verdun campaign had suggested that morale was draining away. In response to the glowing promises emanating from Nivelle this drain had been reversed. French soldiers writing home in March 1917 had used expressions such as 'Victory is smiling on our arms', 'If you could see the enthusiasm of the troops, it's extraordinary' and 'We shall be home in time for the harvest'.

Then came the stark demonstration that Nivelle was offering nothing but the same mixture as before. Revulsion was prompt, and wide-ranging: among politicians, subordinate commanders and rank and file soldiers. The French President and leading members of the government were soon alerted by lower-order members of the military hierarchy that a continuation of operations would achieve nothing more. With dispatch, they set about manoeuvring Nivelle out of the chief post. Effective command was transferred to the one military figure who had stood out against another great offensive: General Pétain.

This turnaround came not a moment too soon. It coincided with the

Operation Alberich

One of the consequences of the Somme campaign was that the Germans undertook a tactical withdrawal to the much stronger Hindenburg position.

The Nivelle Offensive

The limited advances revealed on this map failed to overwhelm the deep German defences and produced profound disappointment, not only in France's governing circles, but among the rank and file soldiers.

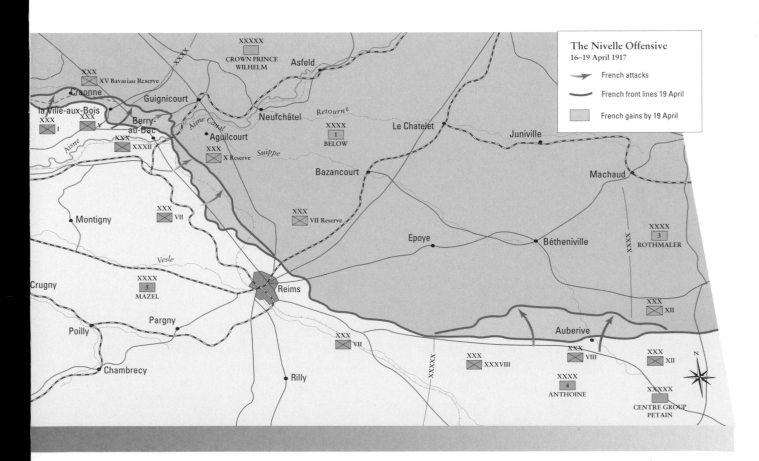

The Nivelle Offensive
16–19 April 1917

French attacks

French front lines 19 April

French gains by 19 April

outbreak among the rank and file soldiers actually engaged in the offensive of a wave of spontaneous acts of 'collective disobedience'. These signalled an ominous message: that the will no longer existed to carry on in the same old way. Among no fewer than sixty-eight of France's 112 divisions – and those, generally, the divisions in the area of the offensive – acts amounting to mutiny occurred, usually in the form of a refusal by troops who had gone into rest to return to the front.

The military hierarchy of France ritually proclaimed these disturbances the work of revolutionaries and agitators. No one behaved as if this was so. The number of executions carried out by the authorities – between fifty and seventy – was remarkably small given the extent of the mutinies. There was an evident reason for this: otherwise, the executions would have had to be intolerably numerous, with possibly yet more dangerous consequences. Overall the considered response at the top was conciliatory. Ordinary soldiers were promised better leave, living conditions and provisions. Above all, the offensive was called off. What was more, it was made evident that – at least in the foreseeable future – no further enterprise on this scale would be contemplated. The French army under Pétain would continue to hold the line. But it would not undertake anything more than some limited set-piece attacks.

What this meant was plain. If the Western Allies were to deliver any major operations against the Germans in the second half of 1917, then it would be up to the British. There was no one else.

THIRD YPRES

The Nivelle catastrophe had a number of implications for Britain. The strategic

initiative, which Lloyd George had endeavoured to unload first on to the Italians and then the French, was back in British hands. And the experiment of transferring command of the British army to a French general was at an end. Haig, if only briefly, had proved the one Allied commander so far that year to enhance his reputation. As a contributor to strategy, Lloyd George had plainly not enhanced his.

Yet it did not follow that Haig was now in a position to dictate Britain's strategy for the remainder of 1917. He had long hankered after a great campaign in Belgium intended to carry British forces all the way from the Ypres salient to the coast. But Britain's political masters were not obliged to endorse this. The military events of 1915 and 1916, and not least Nivelle's operation in the first half of 1917, seemed to point in a different direction. Campaigns with great strategic goals appeared essays in futility. By contrast, limited actions such as the French counter-offensives at Verdun late in 1916 and the British operation at Vimy Ridge in April 1917 were capable of yielding at least a measure of success. That is, set-piece attacks not extending beyond artillery range might not offer victories of Napoleonic proportions, but they could provide worthwhile gains. And they did not threaten to push the troops engaged in them to the breaking point just reached by the French army.

So when Haig, in the aftermath of Nivelle's demise, put forth his ambitious proposal for a campaign out of Ypres, there was no cause for Britain's political leaders to consent. Yet this, in effect, was what happened. No doubt Haig's enhanced stature, and Lloyd George's failed attempt to downgrade him, played a part. But more significant was the fact that Britain's political command, notwithstanding its apparent differences with its military advisers, agreed with

British cavalry in the vicinity of Arras, April 1917. The attempt to employ horsed soldiers on a large scale in the operation confirmed the inappropriateness of cavalry for any major role in a Western Front battle. They were not used again in such numbers, although they continued to be held in readiness for some large enterprise until the last days of the war.

them in a crucial respect. Neither of them believed that Allied strategy should be confined to 'bite and hold' operations of no more than Vimy Ridge dimensions. Lloyd George might not approve of grandiose British offensives on the Western Front. Instead, he embraced ambitious operations by the Italians or the French, or a mighty campaign against the Turks. And once these projected alternatives had failed dismally, or could not be made to appear sensible enough even to merit a try, Lloyd George would conclude (though without compelling reason) that he had no choice but to go along with Haig.

Somewhat by default then, the British army launched its own vast offensive for 1917: the Third Battle of Ypres. Its preliminary phase began promisingly. Haig needed to capture the Messines Ridge to prevent the Germans from observing his main battle preparations. Since 1915, General Sir Herbert Plumer, the commander of the Second Army, had placed a number of great mines – containing all told one million pounds of TNT – under the German front positions. On 7 June these were detonated. At the same time an enormous artillery bombardment was directed against the German batteries. This combination of mine and shell enabled the British infantry to overrun the ridge at moderate cost.

There followed an extended pause while Haig introduced into the area the inexperienced General Sir Hubert Gough to command the main campaign. Haig's action caused operations to lose momentum and gave the Germans time

The battle of Messines. View across the Douve valley, showing shells falling on Messines village.

OPPOSITE: *Preparing for battle at Messines. Australians about to participate in the capture of Messines Ridge study a large-scale model of the area to be attacked. Note the viewing platform towards the top of the picture.*

The first stage of the Third Battle of Ypres. British troops moving over the sodden ground near Pilckem Ridge on 19 August 1917, in one of many failed endeavours.

A derelict Mark IV tank on the Ypres battlefield in September. Even during the productive Plumer phase of the battle, the tank proved of dubious utility on a ploughed-up battlefield.

MESSINES AND
PASSCHENDAELE

This map suggests the vast gulf between what the campaign was intended to do (the capture of the Belgian coast) and what it actually accomplished.

to strengthen their defences in the Ypres salient. So when the attack opened eventually on 31 July, the outcome was ambiguous. The bombardment was of unprecedented ferocity and enabled the infantry to get forward in some areas. But little ground was made towards the crucial Gheluvelt Plateau, which had to be taken if further advances were to be made against the Passchendaele Ridge and towards the Belgian coast. Then a month of heavy rain intervened. This reduced the battlefield to a mire and blighted Gough's every endeavour. By the end of August the line had hardly advanced at all.

Haig responded by moving Gough to flanking operations and giving direction of the main offensive to Plumer. Lengthy preparations were demanded by Plumer, and even then he purposed only an advance of modest proportions. He also had the advantage, to begin with, of fine weather. In battles between late September and early October, he carried out three successful if decidedly limited attacks.

These three battles – Menin Road, Polygon Wood and Broodseinde – were striking examples of the 'bite and hold' method. Enormous amounts of artillery were accumulated – more than had been available to Gough – and barrages fired in front of the advancing infantry to a depth of 1,000 yards. Then, when the objective, which was strictly within artillery range, had been reached, the consolidating troops were protected from the inevitable counter-attack by further barrages. Thereby German forces seeking to recapture lost ground were either swept away or broken up into such small groups that they could not maintain a coherent assault.

The German Command had no answer to these tactics. Certainly, their yield was limited. The amount of ground gained in any attack was no more than 3,000 yards. This mattered because the lateness of the season meant that the fine weather – so necessary for the good artillery observation on which Plumer's method depended – would soon depart.

Indeed, after Plumer's third stroke the rains returned with a vengeance, robbing the artillery of its impact and at the same time turning the battlefield once more into a quagmire. In these circumstances even Plumer's constricted type of operation ceased to be effective. Nevertheless, neither he nor the commander-in-chief chose to call off the campaign. The struggle, from which all large purpose had departed, now developed into a battle for the scarcely significant Passchendaele Ridge. In conditions which beggared the imagination, Haig's army during October and the first half of November clawed its way towards this meagre objective, which when attained constituted a narrow salient vulnerable to flanking shellfire and untenable against a sustained counter-attack. (In three days in 1918, all the ground gained in the Third Ypres offensive was evacuated in the face of a pending German assault.) The whole episode, which the military

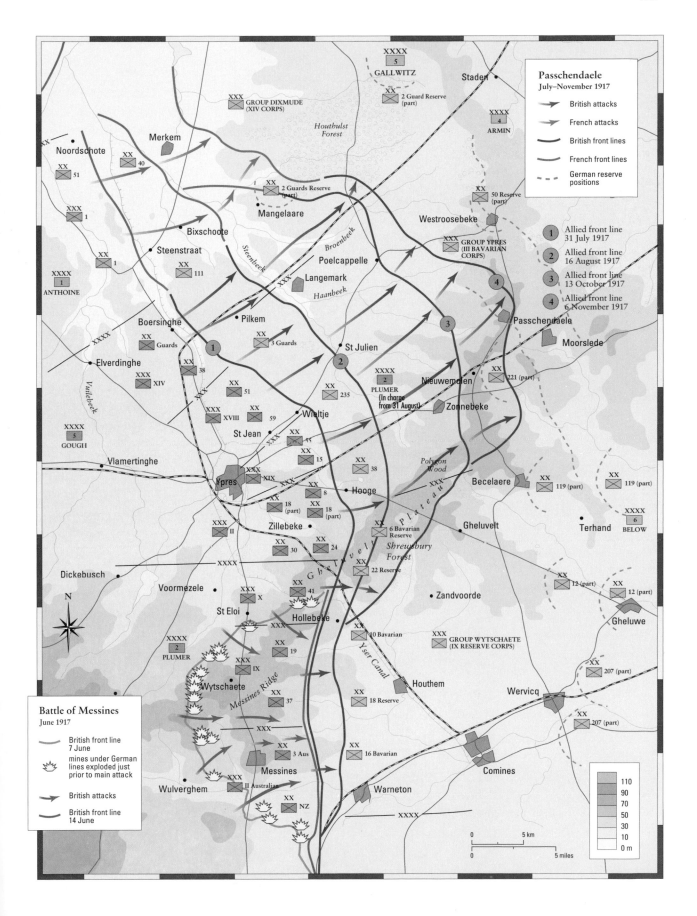

Passchendaele
July–November 1917

→ British attacks
→ French attacks
— British front lines
— French front lines
-- German reserve positions

① Allied front line 31 July 1917
② Allied front line 16 August 1917
③ Allied front line 13 October 1917
④ Allied front line 6 November 1917

XXXX 5 GALLWITZ

XX 2 Guard Reserve (part)

Staden

XXXX 4 ARMIN

XXX GROUP DIXMUDE (XIV CORPS)

Houthulst Forest

Merkem

XX 40

XX 51

Noordschote

XXX 1

XX 1

XXXX 1 ANTHOINE

XX 111

Steenstraat

Bixschoote

Steenbeek

XX 2 Guards Reserve (part)

Mangelaare

Broenbeek

Poelcappelle

Langemark

Haanbeek

XX 50 Reserve (part)

Westroosebeke

XXX GROUP YPRES (III BAVARIAN CORPS)

④

Passchendaele

Moorslede

Boersinghe

Pilkem

XX Guards

XX 3 Guards

①

②

St Julien

XX 38

XX 51

XX 235

XXXX 2 PLUMER (In charge from 31 August)

Nieuwemolen

XX 221 (part)

③

Zonnebeke

Elverdinghe

XXX XIV

XXX XVIII

XX 59

Wieltje

St Jean

XX 15

XX 15

XX 38

Polygon Wood

Becelaere

XX 119 (part)

XX 119 (part)

XXXX 6 BELOW

Terhand

Vlamertinghe

XXX XIX

XX 8

XX 18 (part)

XX 18 (part)

Ypres

Hooge

Gheluveldt Plateau

Gheluvelt

Shrewsbury Forest

XX 6 Bavarian Reserve

XX 22 Reserve

XX 12 (part)

XX 12 (part)

Gheluwe

XXXX 5 GOUGH

XXX II

Zillebeke

XX 30

XX 24

Dickebusch

Zandvoorde

XX 207 (part)

Voormezele

XXX X

XX 41

St Eloi

Hollebeke

Yser Canal

XX 10 Bavarian

XXX GROUP WYTSCHAETE (IX RESERVE CORPS)

Houthem

Wervicq

XXXX 2 PLUMER

XXX IX

XX 19

XX 18 Reserve

XX 207 (part)

Wytschaete

Messines Ridge

XX 37

XX 16 Bavarian

Comines

Wulverghem

XXX II Australian

XX 3 Aus

Messines

Warneton

XX NZ

Battle of Messines
June 1917

— British front line 7 June

💥 mines under German lines exploded just prior to main attack

→ British attacks

— British front line 14 June

0   5 km
0   5 miles

110
90
70
50
30
10
0 m

*The Third Ypres battlefield
at the end of August.*

Not quite the Belgian coast. The 16th Canadian machine-gun company during the last phase of the Third Ypres campaign.

command had senselessly prolonged and which their political chiefs had witnessed disapprovingly but with folded arms, was to the credit of neither generals nor politicians.

CAMBRAI

By the time the Third Ypres campaign was finished, it appeared that operations for the year would cease. But both the British and the German Commands had other ideas.

Haig was looking for action in another area to provide some redemption for the Flanders operation. Further south, in a quiet sector of the Western Front near Cambrai, he observed that the German line was thinly held. As unostentatiously as possible, he assembled troops, artillery, and tanks – but not reserves, which were unavailable owing to the horrendous casualties in Flanders and a recently developed crisis on the Italian front. In the event, the absence of reserves gave him the advantage of surprise, while removing the (strictly hypothetical) possibility of exploiting it.

The choice of a stretch of ground not already devastated by repeated artillery bombardments provided the tanks, this time employed en masse, with their first real opportunity to play a signal role in battle. And of yet greater importance, the introduction of further novel artillery techniques pointed the way to on-going successes in attack, as long as the objectives were suitably restricted. Sound

rangers located the enemy guns with such accuracy that they could be pinpointed on a map, and improved methods of firing ensured that many of the guns so located were destroyed. Moreover, this emerging practice of firing from the map, as a substitute for observing the fall of shot, eliminated the need for a preliminary registration of the guns on enemy targets. This restored an element of surprise, in that the bombardment need only commence as the tanks and the infantry began moving forward.

At Cambrai, this combination of surprise, accurate gunnery and 400 tanks employed over good ground facilitated a swift initial advance at moderate cost to the attackers. These methods could not, however, prevent the Germans from hurrying in reserve forces to shore up the line and halt further progress. What followed was profoundly disappointing for the British. Ten days after the commencement of the attack, German forces – employing storm troops as they

A British tank, accompanied by infantry, in the village of Fontaine during the Cambrai offensive, November 1917.

had done against the Russians at Riga – penetrated the overextended British line and drove Haig's troops back to their starting places. So whatever guidance this action might provide for operations in the following year, for the moment it brought the British Command only fleeting credit, followed by anger and bewilderment.

CAPORETTO

One other action set the war in western Europe briefly ablaze in the closing months of 1917. This was a resounding blow delivered by the German Command against the Italians.

From the start of 1917, Italy's leaders had expressed themselves ready to persist in operations against the Austro-Hungarians as long as the British and French kept the Germans thoroughly occupied. Proposals that they should act in other circumstances, and thereby become a focus for German as well as Habsburg attention, were not acceptable. The events of late 1917 confirmed the wisdom of this view.

Even against just the Austro-Hungarians, Italy's forces in the tenth and eleventh battles of the Isonzo (in May and August) made only meagre progress at heavy cost. By September, Cadorna was informing his allies – to the dismay of

The battle of Cambrai, 20 November 1917. British machine-gunners in a captured German trench.

Lloyd George – that he would be doing no more attacking that year. Then, late in October, his adversaries turned the tables convincingly. The German Command, having triumphed on the Eastern Front and withstood the British in Flanders, diverted forces to Italy where they struck with devastating effect. In the crushing battle of Caporetto, the Italians lost all the territory that they had managed – at a cost of one-third of a million dead and three-quarters of a million wounded – to acquire in the previous thirty months of offensives. In three weeks the armies of the kaiser and the Austrian emperor had advanced an astonishing 80 miles, taken huge numbers of (often unresisting) prisoners, and inflicted wide-ranging demoralization on the Italian forces.

The Italian rout did not prove the sort of collapse witnessed in Russia. Although abrupt changes were made to Italy's government and military command, the social and political structures did not give way. The Italian army rallied on the River Piave, stiffened by an infusion of British and French forces. The Germans, alarmed by events at Cambrai on the Western Front, decided to halt the offensive.

So the threatened loss to the Entente Powers of a second member in the course of 1917 did not come about. Nevertheless, as with France in the aftermath of the Nivelle offensive, a large change had taken place. A member of the Western

Caporetto and its aftermath (1): a German bridging train passes an Austro-Hungarian marching column on a road near Udine (see map on page 61) pursuing retreating Italian forces.

alliance, hitherto resolutely (if unavailingly) on the offensive, had been transformed into a nation shaken by its experiences, licking its wounds, and determined not to mount further large operations unless and until the face of the war had changed dramatically.

GAINS AND LOSSES

The events of 1917 had sent strangely mixed messages to the contestants. Germany had begun the year intending, in all but one respect, to rest on the defensive and marshal its resources. Yet the year had delivered to it the total

elimination of one of its adversaries and heavy rebuffs to two of the others. As a result, Germany's principal ally was relieved of immediate danger, and Germany itself, if it so chose, would be freed from the dilemma of fighting a war on two fronts.

Nevertheless, the events of 1917 had not proved totally to Germany's advantage. Prolonged defensive actions on the Western Front had been costly and draining in both men and matériel. And Germany had sustained a conspicuous setback as a result of its sole attempt to accomplish a major victory in the west. The bid to knock Britain out of the war by a naval campaign against its seaborne trade, and in the process to risk America's abandonment of its neutrality, had yielded the worst possible consequences. Britain's fighting capacity and its ability to sustain its population and commerce were barely diminished. The blockade of Germany at sea remained in place. And the intervention of the United States in the war on the side of the western democracies boded extremely ill for the long-term prospects of the Central Powers.

Seen from the other side, the war in 1917 conveyed this hopeful message to the Entente Powers, and certain other positive aspects as well. Among the latter, we have noted the great array of weaponry which Haig's army now had at its disposal. The fact was that British mobilization of its matériel resources had further gained pace in the course of the year. And in this crucial aspect of modern war Britain was proving a great deal more efficient than the military-dominated economy of Germany.

Caporetto and its aftermath (2). This gives an indication of the type of terrain on which the Austro-German offensive took place. Shown here are troops in a mountain pass.

But offsetting these positive aspects for the Western Allies were several negative features already referred to: the elimination of Russia, the diminution of the fighting capacities of France and Italy, and the squandering of British manpower in the Flanders offensive. The last of these was particularly ominous. Britain, for the moment, was the sole reliable element among the Western European Allies. Yet the Third Battle of Ypres had deprived its armies of the reserves which, if Germany decided to go over to the offensive in the new year, would be necessary at the outset to hold the line and perhaps keep the alliance together.

CHAPTER SIX

1918

'THE GREATEST IMAGE of victory in the whole of the
Great War' (Christopher Moore, Trench Fever, London
1998, p. 216.) The 46th Division following the triumphant
crossing of the St Quentin Canal during the breaking of
the Hindenburg Line. On the left Brigadier J. V. Campbell
VC leans against a wobbly parapet while addressing the
vast concourse of troops, some of whom are still wearing
the life-vests in which they crossed the canal.

1918

CULMINATION

In the course OF discussing the opening battles of the war on the Western Front, reference was made to their colossal scale. So in August and September 1914, eighty or more German divisions were on occasion engaged simultaneously with seventy Allied divisions. Casualties were on a scale which corresponded with these numbers. In just three weeks from 21 August to 12 September, the French lost 330,000 men killed or prisoners – one-sixth of their total dead and prisoners for the entire war.

Nothing in the battles of 1915, 1916 or 1917 approached that intensity. Great campaigns such as Verdun and the Somme in 1916 or the Third Battle of Ypres in 1917 were characterized by daily encounters between a small number of divisions on either side. For example, at the Somme a maximum of eighteen British divisions was engaged at any one time and the average was closer to three or four. All attacks at Third Ypres were delivered by between five and fifteen divisions. This phenomenon was observed by Winston Churchill in the 1920s in his book *The World Crisis*. It has been neglected by historians ever since.

But Churchill noted something else. In 1918 the intensity of battle returned to and then exceeded the levels of 1914. In the first six months of the year when the Germans held the initiative, Ludendorff had at his disposal such reserves of manpower and guns as to enable him to attack with between sixty and eighty divisions against the defending forty to fifty Allied units. When the initiative passed to the Allies in July, the battles became even more extensive. The victory

Preparing for the Ludendorff offensive: a German unit training at Sedan in February 1918.

campaign which began in August was initially a modest affair – sixteen Allied divisions against eight German. By the end of September ninety Allied divisions, supported by quantities of munitions unimaginable in 1914, were engaging a similar number of German divisions. Casualties once more were high.

The sinews of war (1): cannon production at the Bethlehem steelworks, Pennsylvania, USA.

The year 1918, then, witnessed on the Western Front the most extensive fighting of the war. The large battle, often thought to characterize the period 1915–17, was really only a feature of the first and last years of the war.

GERMANY'S OPTIONS

Towards the end of 1917 the German Command surveyed the possibilities for the coming year. Its submarine campaign had clearly failed. And the Americans had joined its opponents. Ludendorff's planners chose to ignore these negatives and to point to more hopeful aspects. They reasoned that the British were, for the moment, exhausted by their efforts at Third Ypres and Cambrai; that the French were incapable of offensive action until at least the summer; and that the Americans had, as yet, few men in the field. In addition, the Bolshevik revolution had taken Russia out of the war, making available (to the extent that Ludendorff

chose) divisions now in the east for the Western Front. Moreover, 100,000 men
could be combed out of industry. Momentarily, these factors would give Germany
a superiority of divisions on the Western Front (192 against 169). For Ludendorff
this was enough. He determined to launch a major offensive in the spring.

An alternative strategy was on offer. It consisted of relinquishing conquests
in the west, straightening the line to save men, building strong defensive positions,
exploiting the resources of conquered Russia and Romania, and leaving the Allies
to do the attacking. This was never considered. Some in Germany might have
doubts about their country's capacity to sustain a major offensive, but they were
not in a position to influence events. The country was now being run by the High
Command.

Where was Ludendorff to strike in the west? The 150 miles of mountainous

region in the south were immediately ruled out. The central French front was tempting but devoid of immediate strategic objectives. In any case Ludendorff did not wish to target the French. Britain was deemed the main enemy and the British army the adversary that must be beaten. One attractive target was the area to the south of Ypres, behind which lay important railway junctions and the not-too-distant Channel ports. But an attack here would engage only a part of the British line. And waiting for the boggy ground to dry might delay operations until the late spring. The ground between Arras and St Quentin would dry earlier, and a massed assault here might roll up the whole of the British sector of the Western Front. So it was between these two points that the German commander determined to launch his offensive.

What means was Ludendorff to employ to achieve his great purpose? In brief, his answer was concentration of force plus tactical innovation. Including troops brought back from the east, he would pack infantry divisions into the chosen front of attack to an extent that would concentrate 750,000 men against 300,000 British defenders. As for artillery, no less than three-quarters of all German guns on the Western Front (some 6,600) would be devoted to the selected area – a figure three times higher than the number of guns available to the British defenders in the sector under attack.

The sinews of war (2): German women stoking a panel oven in a steelworks.

Ludendorff intended to maximize the impact of his infantry by the employment of new tactics. His divisions were divided into shock troops, attack troops and follow-up formations. The most skilled were concentrated into spearhead units called storm troops. They were not to advance in coherent linear formations as of old, but to penetrate deep into the British defences wherever opportunity beckoned, bypassing centres of resistance without waiting for the protection of forces on their flanks. The areas thus bypassed would then be taken out by the follow-up units.

The huge number of guns available to Ludendorff allowed for a short bombardment of incredible ferocity. Brevity, it was hoped, would provide a degree of surprise to the battle. The British rear areas, headquarters and artillery would first be deluged with shells in an attempt to disrupt the command and communication system and to eliminate the main weapons of response. Then the

Exit Russia (1): V. I. Lenin, the Russian revolutionary, who, having ousted the Kerensky government, determined to take Russia out of the war in the belief that other nations would speedily follow.

guns would be turned on to the forward zone in an attempt to stun the defenders just in advance of the main infantry assault.

In some respects these tactics were certainly novel. If the successive British defences could be rapidly breached by this combination of overwhelming firepower and storm troopers, then the German infantry would reach open country and advance rapidly. Nevertheless, there was something reckless and decidedly old-fashioned about this prescription for wide-ranging victory. To achieve the distant objectives that Ludendorff was specifying, and to do so at the required speed, there could be no question of full artillery participation beyond the opening stage. Once the big guns had facilitated the initial rupture, they would soon be left well in the rear. Certainly, Ludendorff enjoined his battery commanders to move their guns forward as swiftly as was practicable. But all experience had confirmed that the rate of movement would not be very swift. And anyway, when the guns did get forward they would need time to establish the whereabouts of their own forces and of the targets they were required to engage.

All this meant that in the aftermath of initial success, the storm troopers would have to exploit success with their own resources. It might be thought that the day had long since departed when a commander on the Western Front would seek to achieve his purposes largely by the actions of his infantry. Yet that, after the opening penetration, was what Ludendorff was contemplating. (As the tactical manual issued to his forces for the offensive stated bluntly, 'the infantry must be warned against too great dependency on the creeping barrage'.) Unless his opponents were so unhinged by initial reverses as to prove incapable of a coherent response, Ludendorff would soon be offering up his last great reserve of manpower to heavy slaughter.

LUDENDORFF'S CHOICE: THE EAST

These plans and preparations for massive operations in the west appeared to require that the German High Command would devote all its attention and resources to that theatre. Yet this was not Ludendorff's response. During 1918, about one million of his men (fifty divisions) were actually retained on the Eastern Front. And for much of this time they were not just holding the line against the Bolsheviks. They were actually undertaking offensive operations.

The need to keep so many troops in the east arose out of an aspect of the

German High Command that has largely escaped attention. Ludendorff and his acolytes were not merely military functionaries. They were also the determiners of their country's foreign policy. That policy was one of aggressive expansionism. So when the Bolsheviks entered into negotiations with the German military, they found themselves confronted with draconian demands.

In the east, German war aims envisaged the dismemberment of much of European Russia and the subordination of the successor states economically and politically to the Reich. So severe were these demands that the Bolsheviks broke off negotiations. Retribution was swift. On 18 February fifty German divisions commenced a three-pronged advance into Russia. This brought Lenin back to the negotiating table. On 3 March the Treaty of Brest-Litovsk was signed.

Its main provisions included the severing from Russia of Finland, the Ukraine, Lithuania, Courland, Livonia and Poland, and the placing of these new states under German tutelage. At a stroke, Russia lost 90 per cent of its coal reserves, 50 per cent of its industry, and 30 per cent of its population.

To enforce this peace, all fifty German divisions still in the east would be retained there. That these troops would be needed soon became obvious. On 10 March the puppet regime set up by the Germans in the Ukraine collapsed. Despite the imminence of the March offensive in the west, Ludendorff ordered that the Ukraine be occupied. Within a week, German forces had entered Kiev

Exit Russia (2): members of the German and Austro-Hungarian delegation receive a Russian delegation at Brest-Litovsk on 15 December 1917 and proceed to make it clear that abject surrender and the vast cession of territory, not a mutual laying down of arms, lies in store for the Bolshevik authorities.

and Kharkov. By April they had reached the Don at Rostov. Instability elsewhere led to the occupation of the Crimea and an advance on the Baku oilfields in the Caucasus. Amazingly, the latter operation continued even beyond Germany's spell of successes on the Western Front. In late July Ludendorff pronounced that Baku oil was a 'vital question'. On 18 August (ten days after he had stated that, in the west, the German army had suffered its blackest day) he ordered that the small British force which had subsequently occupied Baku be expelled. On 10 September, as the Allies were assembling for the climactic assault on the Hindenburg Line, this proposed action at Baku was accomplished by a German

contingent aided by some Russian units (Lenin had been offered a share of the oil by Ludendorff). In late September Ludendorff gathered a team of specialists to proceed to Baku to get the oil flowing. Two days later he announced to the kaiser that the war was lost.

The military implications of all this need to be emphasized. To enforce his programme of aggrandizement against Russia, Ludendorff had to station fifty divisions permanently on the Eastern Front in 1918. Thus one million men were tied to a region where their foes had been thoroughly defeated and from which no military threat now emanated. And this was at a time when the culminating

OVERLEAF: *The limits of the German advance on the Somme, April 1918. The action depicted shows German troops near Villers-Bretonneux, one of them throwing a 'potato-masher' grenade, advancing past the body of a French soldier.*

The dismemberment of Russia: German troops occupy Kiev, the capital of the Ukraine, March 1918.

battles of the war were being fought in the west. Had a less predatory eastern policy been adopted, it has been estimated that the Germans could have moved at least 500,000 troops from Russia to the Western Front. Yet so implacable was Ludendorff's determination to achieve expansion in the east that he carried through his policy to the detriment of his bid for victory in the west. The folly of allowing the military to dominate all aspects of policy in Germany could not have been better illustrated.

LUDENDORFF'S CHOICE: THE WEST

In the west, Ludendorff's offensive opened on 21 March. It fell on the weak defences of the British Fifth Army and the rather stronger defences of the Third Army between St Quentin and Arras. The German artillery was so superior numerically to that of the British that in the south the storm troopers speedily broke out into open country. In a week they advanced 40 miles on a 50-mile front – a feat without parallel in the stalemate period – and were approaching the important rail junction of Amiens. All this was a confirmation of the value of the storm troop method in the opening phase of a battle against rudimentary defences. But soon the crucial shortcomings in the method revealed themselves.

OPERATION MICHAEL

The first phase of the Ludendorff offensive, March–April 1918. This was perhaps the most dangerous for the Allies of all the German offensives. However, the Germans failed to make significant advances against the British Third Army in the north, or capture the important rail junction of Amiens in the south.

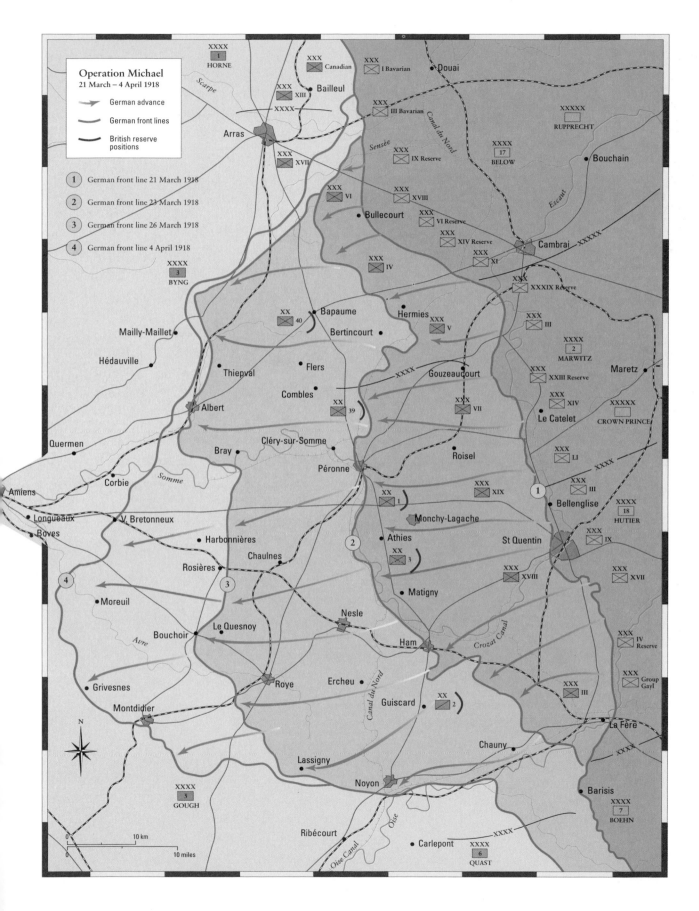

Operation Michael
21 March – 4 April 1918

→ German advance

〜 German front lines

〜 British reserve positions

① German front line 21 March 1918

② German front line 23 March 1918

③ German front line 26 March 1918

④ German front line 4 April 1918

XXXX
1
HORNE

XXX Canadian

XXX I Bavarian

Douai

XXX XIII

Bailleul

XXXX

XXX III Bavarian

Scarpe

Sensée

XXXXX
RUPPRECHT

Canal du Nord

XXX IX Reserve

XXXX
17
BELOW

Bouchain

Arras

XXX XVII

XXX VI

XXX XVIII

XXX VI Reserve

Escant

XXX XIV Reserve

Bullecourt

XXX XI

Cambrai XXXXX

XXX IV

XXX XXXIX Reserve

XXXX
3
BYNG

Mailly-Maillet

XX 40

Bapaume

Hermies

XXX V

XXX III

Bertincourt

XXXX
2
MARWITZ

Hédauville

Thiepval

Flers

Gouzeaucourt

XXXX

XXX XXIII Reserve

Maretz

Combles

Albert

XX 39

XXX VII

XXX XIV

Le Catelet

XXXXX
CROWN PRINCE

Quermen

Bray

Cléry-sur-Somme

Roisel

XXX LI

Somme

Péronne

XXX XIX

Corbie

XX 1

XXXX

Amiens

Monchy-Lagache

XXX III

XXX IX

Longueaux

②

Athies

XXXX
18
HUTIER

Boves

V. Bretonneux

Harbonnières

XX 3

St Quentin

XXX XVII

Chaulnes

XXX XVIII

Rosières

③

Matigny

④

Moreuil

Bouchoir

Le Quesnoy

Avre

Ham

Crozat Canal

XXX IV Reserve

Grivesnes

Roye

Ercheu

XX 2

Guiscard

XXX III

XXX Group Gayl

Montdidier

Canal du Nord

La Fère

N

Chauny

Lassigny

Noyon

Barisis

XXXX
5
GOUGH

XXXX
7
BOEHN

0   10 km

0   10 miles

Ribécourt

Oise

Carlepont

XXXX
6
QUAST

XXXX

Oise Canal

①

Bellenglise

The Ludendorff offensive, phase one: an improvised French and British defensive line in open country, evidence that the Germans had broken through the Allied trench positions.

The German troops were approaching exhaustion. Casualties, especially in the élite storm troop formations, had been heavy. The great mass of the artillery was still struggling to get forward. Increasingly, therefore, the infantry had only their light weapons to rely upon for fire support. On the other side of the line, the British were rushing reserves and guns forward by rail. These came from the French sector, from the unattacked portion of the British front, and even from Britain itself. The inevitable consequence was that the German onrush was brought conclusively to a halt.

Thwarted in Picardy, Ludendorff turned north, where the ground had now dried. On 9 April he attacked to the south of the Ypres salient. His method once

Actions of the Somme crossings, March 1918: support troops of the British 20th Division lining the edge of a road.

The Ludendorff offensive, phase one: pipers leading a column of British troops in orderly retirement, 25 March 1918.

The Ludendorff offensive, phase two: 'backs to the wall'. British soldiers behind an improvised road-block during the battle of the Lys, 29 April 1918.

more yielded immediate gains, especially on the front of two reluctant Portuguese divisions. Here too, however, and for the same reasons as already noted, the front eventually stabilized. No strategic goals had been achieved. The British army, declared by Ludendorff to be his deadliest foe, was battered but still intact.

Also firmly in place was the main union between the great Western Allies. On 26 March the British and French political and military leadership had met at Doullens. An earlier suggestion by Pétain that the British and French armies should separate so as to cover respectively the Channel ports and Paris was swept aside. So, anyway on this matter, was Pétain. General Ferdinand Foch was placed in overall charge of all Allied armies on the Western Front. If this did not have the military significance that some commentators have claimed, it was certainly evidence that the alliance would hold together and fight on united.

Unity of command. On 26 March General Foch was placed in overall command of French and British armies on the Western Front.

Meanwhile, Ludendorff had to decide on what to do next. In a convoluted argument, he declared that, while Britain was still the main foe, he must now attack the French. His justification was that it had been French reserves that had saved the British in March and April. These reserves must now be eliminated or pinned to the French sector. Then he could once more turn his attention to the British. In fact, this reasoning was merely an elaborate justification for the course that had been forced upon him by events. If he was to mount a further offensive in reasonable time it must be to the south. Returning to the Somme or pressing on in Flanders held no attraction, and the only other options available on the British sector – the Lens–Loos industrial wastelands or the formidable heights of Vimy Ridge – were not appealing.

So Ludendorff chose to attack the French on the Chemin des Dames, an area which had seen no fighting since October 1917. His forty-division assault, backed by 3,500 guns, was launched on 27 May. His methods again won immediate gains, helped on this occasion by the obtuse policy of the French army commander. Incomprehensibly, General Denis Duchêne crowded most of his troops into the forward zone and thus rendered them easy targets for the German gunners. Moreover, on one sector of the front all that opposed the Germans were the shattered remnants of eight British divisions sent to the area to recuperate after their devastation in the March offensive. In three days, the Germans advanced 40 miles. The following day saw them once more on the Marne and in sight of Paris.

OPERATION GEORGETTE

The Ludendorff offensive, phase two: the German thrust towards the Belgian coast, April 1918.

OPERATION BLÜCHER–YORCK

The Ludendorff offensive, phase three: the German drive towards Paris.

Then, as French reserves began arriving, the pace of the German advance slackened. This might have been of little concern to Ludendorff. He had clearly achieved his purpose of eliminating many French divisions and attracting others away from the British front. But his objective was changing. Paris now beckoned. It lay just 40 miles from his foremost troops. The opportunity seemed too good to miss. Reinforcements were hurried from Flanders (the proposed area of his renewed offensive against the British) to assist in the advance on Paris. Yet the opportunity proved not to be there. His new formations could gain no ground against a re-established French defence. Their only contribution was to an increased German casualty list.

Despite this failure, Ludendorff decided to persist in this region. Paris, rather than the British army, was now his great objective. But the appearance of

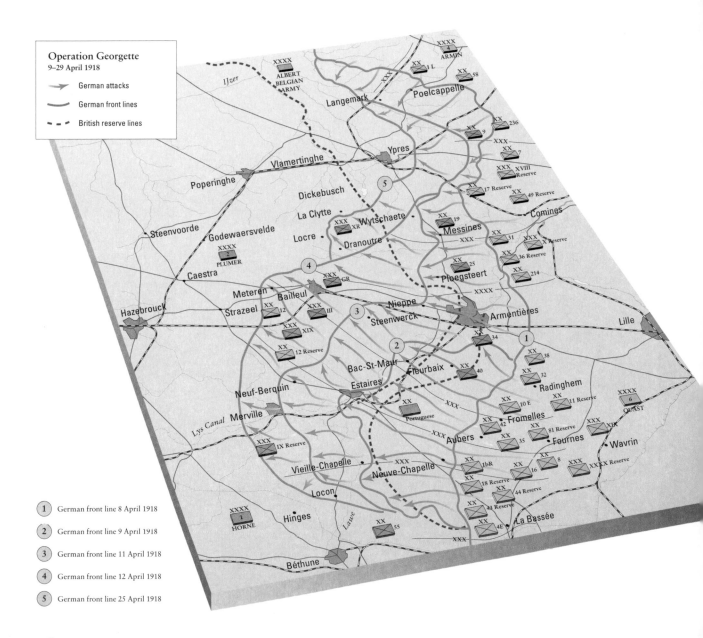

Operation Georgette
9–29 April 1918

→ German attacks

⌒ German front lines

- - - British reserve lines

① German front line 8 April 1918

② German front line 9 April 1918

③ German front line 11 April 1918

④ German front line 12 April 1918

⑤ German front line 25 April 1918

*The Ludendorff offensive,
phase three: the limits of the
German advance against the
French, July 1918.
Ludendorff's troops near
Château Thierry, where they
will encounter American
troops for the first time in
the Great War.*

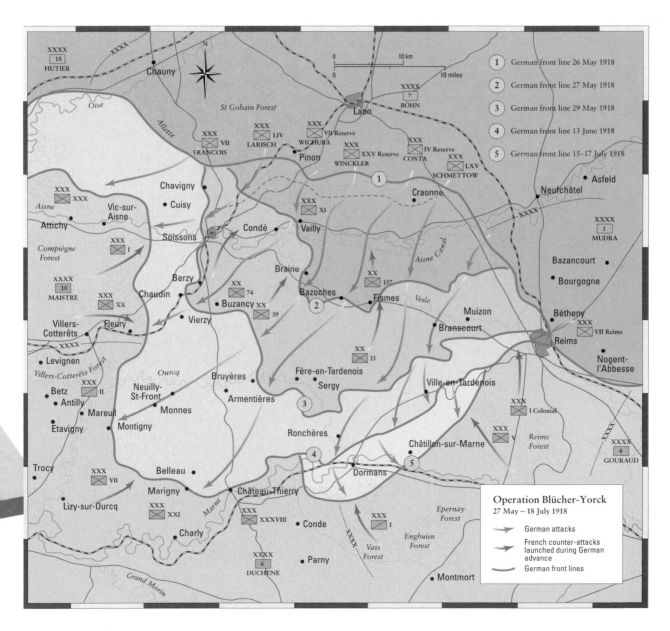

①	German front line 26 May 1918
②	German front line 27 May 1918
③	German front line 29 May 1918
④	German front line 13 June 1918
⑤	German front line 15–17 July 1918

Operation Blücher-Yorck
27 May – 18 July 1918

→ German attacks

→ French counter-attacks
launched during German
advance

— German front lines

American troops on the Marne indicated that the balance of resources was moving decisively against him. And the need to act in haste deprived his armies of the careful preparations that had characterized his offensives so far. So the next two operations launched against the French – one to the north of the Chemin des Dames on 9 June, the other on the Marne on 15 July – lacked thorough planning and the element of surprise. Both were halted by the artillery and reserve divisions brought forward by the thoroughly alerted French Command. Nevertheless, Ludendorff declared himself satisfied. In a reversion to his original plan, he noted that French reserves had now been drawn into the southern battle. So he hurried to Flanders to prepare a new offensive against the British. But just as he reached his headquarters there, dramatic events began to unfold. His adversaries were moving to the attack.

MIDWAY

How can we sum up the German effort between March and July 1918? Considerable amounts of territory had certainly been overrun, especially in the area of the old Somme battlefield and between the Aisne and the Marne. In five months Ludendorff had succeeded in capturing more territory than the Allies had managed in three years.

The Ludendorff offensive, phase four: troops of the German Eighteenth Army between Montdidier and Noyon, June 1918.

Yet none of the ground taken was of vital strategic or even tactical importance. Major rail junctions, such as Amiens and Hazebrouck, had evaded Ludendorff's grasp. Moreover, the great bulges created in the Germans' line by their advances meant that in July Ludendorff was having to hold twice the length of front he had held in March. And he was required to do this with far fewer men. German casualties during these months were probably about one million. The Allies too had suffered grievously – they had probably lost about 900,000 men. But the Germans were less able to bear their losses. Their manpower pool was smaller than that of Britain and France combined, and they did not have the prospect of limitless American reinforcements. Furthermore, despite their successes, they had not yet found a formula for victory. Storm troop tactics had proved able to break the enemy front, but not to prevent fresh lines from being formed ahead of the German advance. The end result of Ludendorff's efforts, therefore, was to deplete his army and to place much of it in poorly defended salients. That was all.

THE GREAT REVERSAL

The Allied counter-offensive began on the Marne. On 18 July the French Tenth and Sixth Armies, accompanied by 750 tanks, fell on the flank of the large salient

German 6-inch howitzers were employed in the open during their offensives.

created in the battles of May and June. The Germans, outnumbered three to one and with many of their divisions second-class, ordered an immediate withdrawal. In the days that followed, the battle was extended to encompass both flanks of the salient. Under this pressure the German retreat gathered pace and by 7 August they were back on the Aisne. The offensive also spelled the end of Ludendorff's Flanders ambitions. That proposed operation was cancelled and he returned south to the crisis area.

By the time he arrived, something akin to stalemate appeared to have returned to the battlefield. The French offensive had ground to a halt. Almost the entire French tank force had been expended, the troops were exhausted, and German reserves had stabilized the front. Foch's operation, in short, had gained some ground but had run out of steam for exactly the same reasons as had the German attacks. Nor were the French capable of a follow-up operation in the near future. That task, as Foch made clear, must lie with the British.

As it happened, at that very moment just such an offensive was being planned in the area of the British Fourth Army to the south of the Somme. The proposed attack had the modest aim of freeing the rail junction of Amiens from German long-range artillery fire. Yet although the intentions were modest, the means of execution were not. This might seem surprising. During the great retreats in the first half of the year, British losses in weaponry had been enormous. Over 1,000 artillery pieces had been lost, along with innumerable machine-guns and lighter weapons. Nor had Haig's army yet made up for its losses in manpower. By June 1918 a British division was down to about half the number of riflemen of 1916.

Nevertheless, there was a factor on the British side that more than compensated for these deficiencies. It lay in the prodigious accomplishments of the nation's munitions industries. By July the Ministry of Munitions had resupplied the army to such an extent that it now possessed more artillery than it had disposed of on 21 March. In addition, British industry was delivering increasing quantities of Lewis guns, machine-guns, trench mortars, smoke, gas and high explosive shells, as well as aircraft and tanks of superior quality. What this meant in terms of the battlefield was that the smaller number of infantry in a 1918 division could deliver much more firepower than its 1916 equivalent. And in this war it was firepower rather than infantry numbers that decided battles.

For the enemy which faced the Fourth Army, there were no such compensating factors. By almost every measure of economic activity, Germany was in steep decline by 1918. Taking 1913 as the benchmark, industrial output had diminished by a third by 1918.

This had been brought about by a number of factors. The blockade had deprived Germany of vital strategic materials such as cotton and nitrates, the shortage of which made the manufacture of munitions more difficult. More importantly, since August 1916 almost the entire German economy had been taken over by the military. This circumstance had been a consequence of the Battle of the Somme where, as noted, the Germans had been shocked by the superiority in weapons of the British. The result, however, had been a series of poorly thought out crash programmes which hardly served the purposes of boosting munitions production. For example, so many shell factories were constructed as to leave an actual shortage of steel for the manufacture of shells. At the same time, the rail system of the Reich had begun to collapse because vital activities such as the maintenance and replacement of rolling stock had been seriously neglected by the military. One consequence was that coal could not reach the factories for want of adequate transportation.

Overall, although in the short run German munitions production increased for a while, it did so in a most inefficient way and at the expense of the civilian economy. This last aspect applied particularly to the food distribution network, whose decrepit state and monopolization by the military left some areas of Germany short of food, even though food stocks overall were sufficient. It was the matter of food availability that, more than any other factor, led to unrest on the home front: strikes, undernourishment and the collapse of real wages, resulting in further reductions in the output of the war-related industries.

On the battlefield the overall decline was becoming starkly apparent. After the failed offensives of the first half of 1918, the German High Command also had to reduce the strength of its divisions. But it could expect no firepower increment from Germany's declining war industries. So when the Allies went over to the offensive, the equipment they captured or destroyed would not be replaced.

The Advance to Victory. A dump of German guns captured by the British Fourth Army at the battle of Amiens, 8–12 August 1918. The German army faced increasing difficulty in securing replacements for captured and destroyed weapons.

The gap between the well-supplied Allies and their declining opponents became evident as soon as the Fourth Army began its assault in front of Amiens on 8 August. By the end of the day, the Australian, Canadian and British units had advanced 8 miles on a 9-mile front, captured 400 guns, and inflicted 27,000 casualties. Their losses amounted to only 9,000 men.

The key to this success lay in the weapons system which the British had been developing since Cambrai. Foremost among the elements of this system was the artillery. At the battle of Amiens, the British guns outnumbered the German by more than two to one. More importantly, by 1918 the British, using sophisticated location devices such as sound-ranging, could pinpoint enemy guns without prior attempts to find the range. Given that preliminary ranging had almost always revealed the intention to attack, these developments in the employment of artillery helped to restore surprise to the battlefield. So when the bombardment

ALLIED ADVANCE,
8–25 AUGUST 1918

*Days 1 to 18 of the Allies'
Hundred Days of Victory.
This was the first major
defeat inflicted on the
Germans by the British
army in 1918.*

came down on the morning of 8 August, most German guns were blanketed or destroyed. At one stroke, the main impediment to the attacking infantry and to the progress of the tanks was eliminated.

The other great opponent of the infantry, the machine-gun, was dealt with by further elements of the weapons system. The creeping barrage of high explosive shells kept down the heads of the enemy machine-gunners and other defenders until they could be set upon by the troops advancing just behind the barrage.

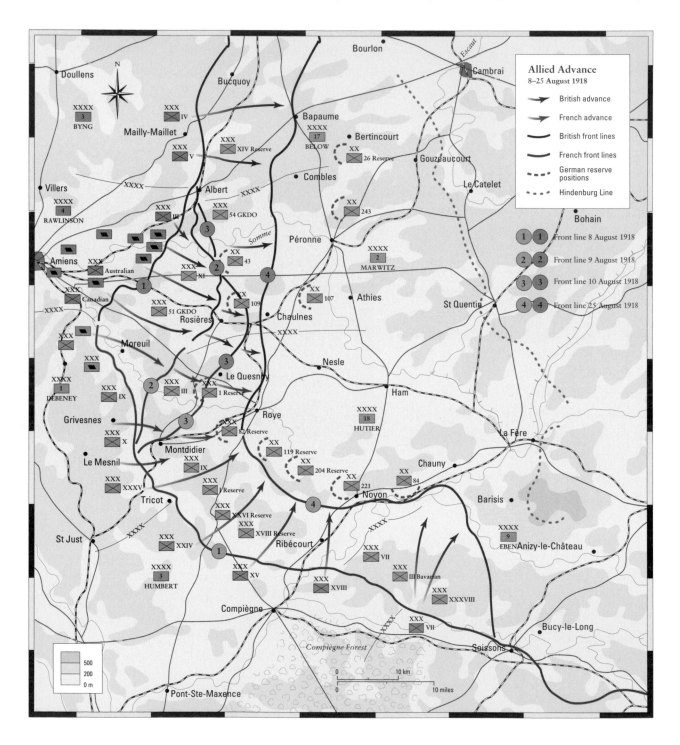

Those missed by the shelling were eliminated by outflanking movements of troops equipped with mortars, Lewis guns and rifle-grenades. Finally, the tanks – 400 of the more reliable Mark V variety – unimpeded by hostile artillery, helped keep down casualties by themselves dealing with pockets of resistance and by causing, in some instances, enemy troops to flee the battlefield.

It is important to note the widespread applicability of this weapons system. Once the British, by employing a combination of big guns, mortars, machine-guns, tanks and aircraft, had devised a method of dominating German artillery and trench defences, they were in a position to get their troops forward at least as far as the distance that a high explosive shell could travel. Moreover, it was not important in these circumstances whether enemy morale was secure or waning. At Amiens, German troops of high morale were overrun just as thoroughly as those whose devotion to combat was less than robust.

But a major test of the new methods of conducting an offensive remained. After all, the Germans' defences at Amiens – an area they had but recently overrun and to which their High Command had since paid little attention – were rudimentary. But well behind the front stood the altogether more formidable Hindenburg Line, which in the aftermath of the Battle of the Somme in 1916 the German Command had developed into a sophisticated defensive system. Would the British army's method of attack also prove capable of overcoming this kind of defence in depth?

This question did not arise immediately. In the aftermath of the spectacular success of 8 August, the Fourth Army's offensive declined in impact. Operations on 9, 10 and 11 August produced increasingly uncoordinated attacks with steadily diminishing returns. The painful question presented itself: would the British army yet again, by attempting to convert limited success into something much greater, manage to replace moderate accomplishment by purposeless bloodletting?

The answer, at last, was to be in the negative. Haig wanted to press on, as always. But he encountered firm resistance from lower-order commanders, not

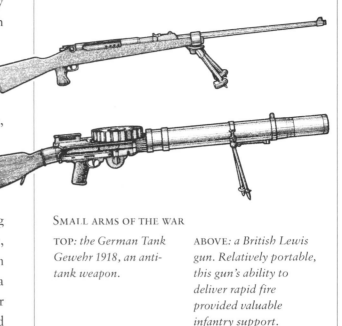

SMALL ARMS OF THE WAR

TOP: *the German Tank Gewehr 1918, an anti-tank weapon.*

ABOVE: *a British Lewis gun. Relatively portable, this gun's ability to deliver rapid fire provided valuable infantry support.*

A British observation post for artillery in Belgium during the Allied advance.

British troops advancing in
Belgium accompanied by
light tanks. On the right of
the picture, one of the tanks
has gone up in flames, with
the body of a soldier lying
nearby.

MEUSE–ARGONNE
OFFENSIVE

*The map opposite shows
Franco-American
operations in the
Meuse–Argonne region,
September–November
1918.*

US FORCES ON THE
WESTERN FRONT

*This map show the sites of
the major battles in which
the US army participated.*

**United States involvement
in the First World War**
1917–18

→ major US attacks

— German front lines

— French front lines

— US front lines

① German front line July 1918

② Belgian front line 11 November 1918

③ British front line 11 November 1918

④ French front line 11 November 1918

⑤ US front line 11 November 1918

US offensive

☐1 18 July – 6 August 1918
Aisne–Marne

☐2 12–16 September 1918
St Mihiel

☐3 26 September – 11 November 1918
Meuse–Argonne

least among them General Sir Arthur Currie, the chief of the Canadian forces (who could, if need be, call upon significant backing from his own government). Haig conveyed these views to Foch, supposedly the supreme commander. Foch insisted that Haig disregard the objections of his subordinates and continue hammering away. At this point the British commander-in-chief underwent something of a change. He told Foch that he would mount attacks on other parts of the line, but he would not continue this particular offensive. And he warned that if Foch ordered him to do otherwise, he would appeal to his own government. Foch came to heel. Thus, at last, the Allied Command in the west, under pressure from subordinate leaders guided by the clear logic of what constituted productive operations, resorted to a manner of proceeding, which, if it would yield no massive victories, would produce a succession of mounting triumphs.

So, on 12 August the battle of Amiens was closed down. Operations by the other British armies swiftly followed, using similar methods to those employed on 8 August. First one British army, then another, moved to the attack. When the offensive lagged on one section of the front it was taken up on another. By these means the Germans were kept wrong-footed, and were prevented from sending all their reinforcements to one particular section of the line.

By early September, then, almost the entire British front was moving. Either the Germans were being outfought, as the seizure of the tactically important Mont St Quentin by an Australian division demonstrated, or they were being forced to withdraw to maintain a coherent front, as was the case in Flanders. By the middle of the month the outlying areas of the mighty Hindenburg Line had been reached by Allied troops.

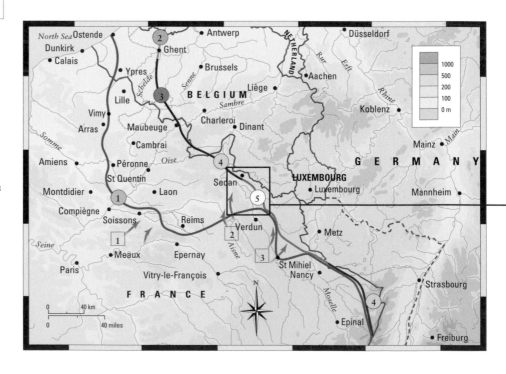

The next series of battles saw the climax of the war. Between 26 and 29 September, the fifty divisions of the British army, along with the Belgians in the north and formidable contingents of French and Americans in the more southerly sector, assailed the Germans along almost the entire length of the Western Front.

The Americans and French moved first. On 26 September they attacked in the Meuse–Argonne region. So far the Americans had played only a minor role in Western Front fighting. They had assisted the French defence on the Marne in the middle of the year and the Australians in a minor operation at Hamel in July.

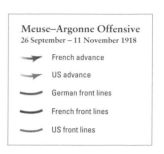

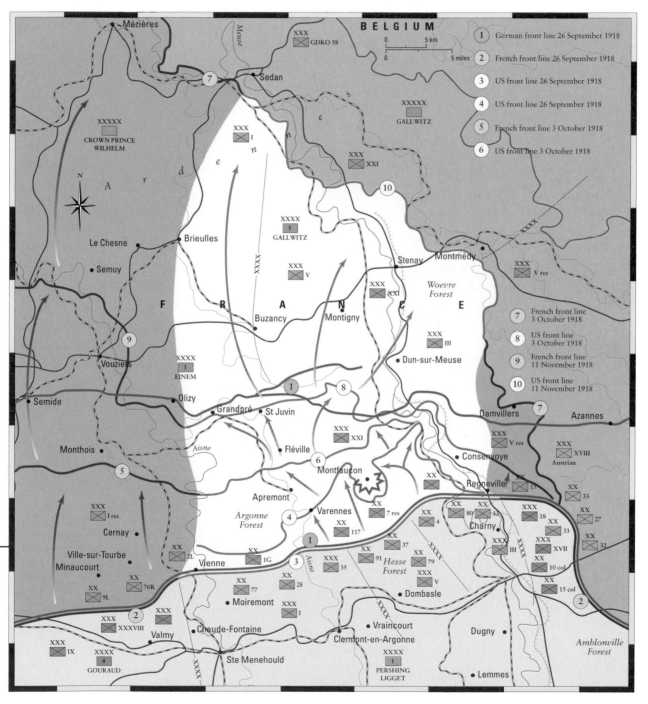

Then on 12 September they had cleared the St Mihiel salient to the south of Verdun. The Meuse–Argonne was a much larger affair involving fifteen American and twenty-two French divisions. The area of operation was difficult. The devastated country to the north of Verdun was devoid of good communications, the Argonne forest was spread across a tangled array of rivers and ridges, and the deployment of massed tanks was impossible. In addition, the American troops lacked experience and the Germans possessed ample time to build a formidable defence. In these circumstances the operation made only slow progress. By October the French and the Americans had gained about 15 miles and were approaching the main German defensive position, the Kriemheld Line, but by that time the operation had been overshadowed by events further north. Nevertheless, this Franco-American attack succeeded in pinning down thirty-six German divisions in the southern area, and that in itself was of considerable assistance to the other armies.

General John Pershing, commander of the American Expeditionary Force.

The operations to the north, which commenced on 27 September, involved five British and two French armies, the Belgian army, and two divisions of American troops serving with the British. The Hindenburg defences were formidable. In places they were 3 miles deep, with protecting wire and concrete machine-gun posts. In some sections they were aligned to incorporate steep-banked canals. The key sector lay between Epehy and St Quentin, in the area of the British Fourth Army. If this position could be broken, the entire line to the north would be turned. But it was here that the defences were at their strongest. The main element was the St Quentin Canal, which had 50 feet of steeply sloping banks, and water or mud to a depth of 6 feet. It was an insuperable barrier to tanks and a considerable obstacle for infantry. Just to the north, the canal ran through a tunnel which should have made the going easier but consequently encompassed the greatest depth of defence.

Against these obstacles there was no question of employing surprise, as had been possible at Amiens. A long bombardment on Somme lines was essential to destroy enough wire and machine-gun posts to allow the passage of the infantry. Also, in contrast to Amiens, tanks could play only a minor role. They were useless against the canal and of dubious utility against the tangle of defences in the covered-over tunnel area.

Despite these problems, several factors were to the advantage of the British.

First, they had captured plans of a section of these defences revealing every machine-gun post, artillery position, trench and wire entanglement. Second, the British were able to employ to ever greater effect the method of maximizing artillery fire which had been so successfully applied at Amiens. And third, British industry had supplied the artillery with high explosive shells in unprecedented numbers.

The action of 29 September revealed the potency of these factors. The counter-batteries proved just as effective as on 8 August. As a result, most German guns had been neutralized by zero hour and played little role in the ensuing battle.

Certainly, not all aspects of the attack went well. In the northern sector, where theoretically the tanks could be employed against the tunnel, the powerful defences held up the American and Australian attackers and thus deprived them of the supporting barrage. Without this protection the infantry suffered prohibitive casualties and were brought to a halt. The tanks could make little

Approaching the war's climax: British troops passing through the remains of German wire entanglements near Bellicourt, 4 October 1918.

A section of British Mark V tanks moving to the attack on the Hindenburg Line. Note the structures on top of the tanks which were dropped into trenches, thus allowing the tanks to cross.

progress against deep entrenchments. The entire attack across the tunnel stalled.

But events further south, where the canal defences happened to be at their strongest, redeemed this setback. In the aftermath of a devastating artillery bombardment, an obscure British division (46th North Midland) crossed the canal and pushed on to breach the Hindenburg defences on a 3-mile front. Thereby they outflanked the Germans holding up the Australians and Americans, and enabled the attack to proceed along the whole front. The key to this success should be carefully noted. Because of the evident difficulties which would be met in attempting the crossing of a canal, the Fourth Army had concentrated most of its artillery in the Midlanders' area. As mentioned, the counter-batteries early on had eliminated the distant German guns. The remainder of Fourth Army's artillery, employing a huge volume of shells, overwhelmed the more immediate defences. Some statistics illustrate this proceeding. For each minute of the attack, 126 shells from the field guns alone fell on every 500 yards of German trench. And this intensity was maintained for the entire eight hours of the attack. That is, in the advance from the near bank of the canal to their final objective, these

infantrymen on any 500 yards of front were supported by 50,000 shells. No defences could withstand this onslaught. It was hardly surprising that the defenders, irrespective of their morale, were killed, stunned or too cowed to offer protracted resistance.

These events, with local variations, were repeated in the areas of the First and Third British Armies. By 5 October the Allies were through the entire Hindenburg system and into open country.

It was clear what these operations signified. The British had now developed methods of overwhelming the most powerful defensive system at relatively modest cost. There was of course a severe limitation: no advance could be pushed beyond the protection of the covering artillery. By the end of September even Haig, though at times reluctantly, had come to see the wisdom of this 'bite and hold' approach, at least as a prelude to the still-anticipated climactic battle for which the cavalry remained in readiness.

So in essence, from early October until 11 November, the Allies continued to make a series of steady, if unspectacular, advances along their entire front,

OVERLEAF: *A patrol of the Lancashire Fusiliers enter the ruined city of Cambrai. Cambrai had been a British objective since the Somme campaign two and a half years earlier.*

The last days: British troops (with juvenile assistance) entering Lille, occupied by the Germans since 1914 and now abandoned without a fight.

ADVANCE TO VICTORY, AUTUMN 1918

This map shows the northern section of the Western Front. Although advances were made by Franco-American forces further south, it was in the area depicted on the map that the classic defeats were inflicted on the German army.

Allied Advance

September–November 1918

area retaken by Allies to 11 Nov. 1918

still occupied by German forces at Armistice

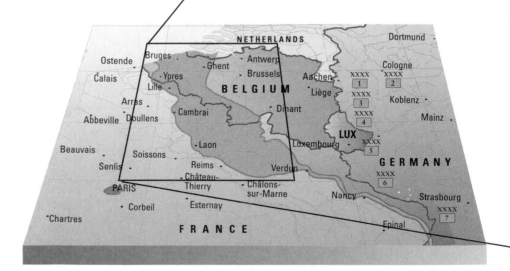

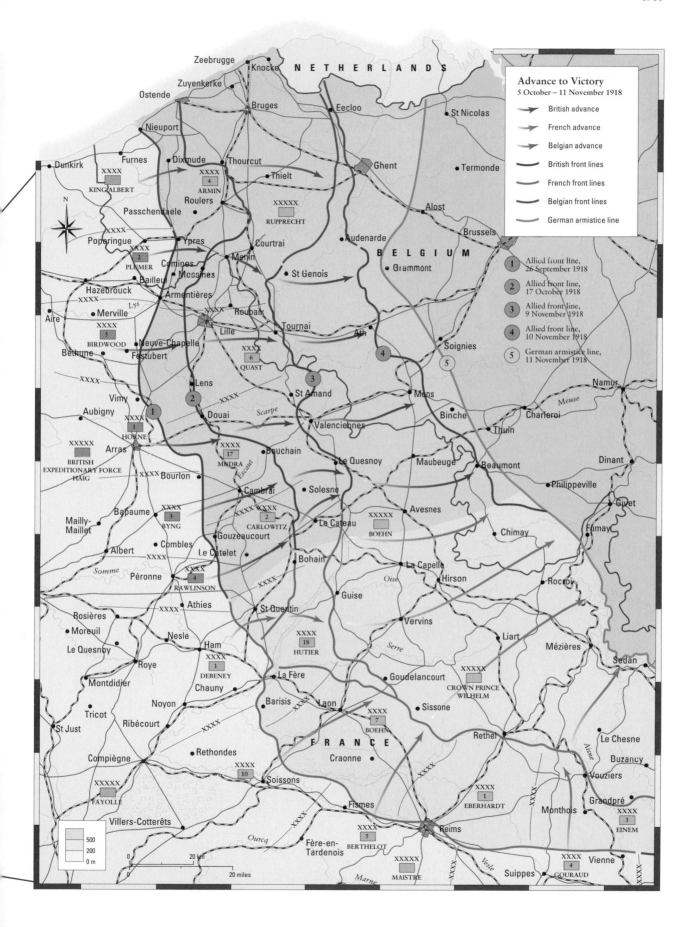

Advance to Victory
5 October – 11 November 1918

→ British advance
→ French advance
→ Belgian advance
━ British front lines
━ French front lines
━ Belgian front lines
━ German armistice line

① Allied front line, 26 September 1918
② Allied front line, 17 October 1918
③ Allied front line, 9 November 1918
④ Allied front line, 10 November 1918
⑤ German armistice line, 11 November 1918

NETHERLANDS

Zeebrugge
Knocke
Ostende
Zuyenkerke
Eecloo
Bruges
St Nicolas
Nieuport
Ghent
Termonde
Dunkirk
Furnes
Dixmude
Thourcut
XXXX 4 ARMIN
Thielt
Alost
KING ALBERT
Roulers
XXXXX RUPPRECHT
Brussels
Passchendaele
Ypres
Courtrai
Audenarde
BELGIUM
Poperingue
Menin
Grammont
XXXX 2 PLUMER
Comines
Mossines
St Genois
Hazebrouck
Bailleul
Armentières
Roubaix
Ath
Soignies
Aire
Merville
Lys
XXXX
Lille
Tournai
⑤
Bethune
XXXX 5 BIRDWOOD
Neuve-Chapelle
Festubert
XXXX 6 QUAST
④
Namur
Meuse
Lens
St Amand
Mons
Charleroi
Vimy
XXXX
②
③
Binche
Thuin
Aubigny
Douai
Scarpe
Valenciennes
Maubeuge
Beaumont
Dinant
XXXX 1 HORNE
①
XXXX 17 MUDRA
Bouchain
Le Quesnoy
Philippeville
Arras
Givet
BRITISH EXPEDITIONARY FORCE HAIG
XXXX Bourlon
Cambrai
Solesne
Avesnes
Fumay
Bapaume
XXXX 3 BYNG
XXXX XXXX 2 CARLOWITZ
Le Cateau
XXXXX BOEHN
Chimay
Mailly-Maillet
Gouzeaucourt
Combles
Le Catelet
Bohain
La Capelle
Hirson
Rocroi
Albert
Somme
Péronne
XXXX 4 RAWLINSON
Oise
Guise
Vervins
Liart
Mézières
Rosières
Athies
XXXX
St Quentin
Serre
Sedan
Moreuil
Nesle
XXXX 18 HUTIER
Le Quesnoy
Ham
XXXX 1 DEBENEY
La Fère
Goudelancourt
XXXXX CROWN PRINCE WILHELM
Le Chesne
Roye
Montdidier
Chauny
Laon
Sissone
Rethel
Buzancy
Tricot
Noyon
Barisis
XXXX 7 BOEHN
Vouziers
St Just
Ribécourt
FRANCE
Craonne
Aisne
Grandpré
Compiègne
Rethondes
XXXX 10
Soissons
Fismes
XXXX 1 EBERHARDT
Monthois
XXXX 3 EINEM
XXXXX FAYOLLE
Villers-Cotterêts
Ourcq
Fère-en-Tardenois
XXXX 5 BERTHELOT
Reims
Vesle
Vienne
XXXXX MAISTRE
Marne
Suippes
XXXX 4 GOURAUD

500
200
0 m

0   20 km
0   20 miles

pausing to consolidate at times, then advancing once more. By 17 October the Germans had lost the line of the River Selle, and in early November the Scheldt and then the Sambre. Their increasingly disorganized armies could do little but accelerate their retreat.

Ludendorff, in a lucid moment on 28 September, had realized that he had no answer to this onslaught. He recommended making peace. That he then changed his mind did not signify. The newly appointed civilian government in Germany disregarded his latest about-turn and sought to initiate armistice negotiations. This precipitated his resignation. On the night of 7/8 November a German delegation crossed the French line to commence discussions. With widespread strikes at home, mutiny in the fleet, and revolution threatening in a number of regions, the delegation had little choice but to sign. So at 11 a.m. on 11 November the war on the Western Front came to an end.

OTHER FRONTS

Meanwhile, Germany's setbacks in France and Flanders had precipitated a collapse of its allies.

In Italy the disintegration of the Central Powers had been hastened by a botched Austro-Hungarian attack across the Piave in mid-June. It was followed in due course by an Italian counter-attack. What became clear was that the Habsburg army was no longer in a condition even to hold the line. It was short of transport, food and supplies and equipment of all kinds. All sense of purpose now departed. In the three months after the Piave, 200,000 soldiers deserted. By the end of September this number had doubled. What was happening was that the 'nationalities', which up to now had fought so hard against the Italians, were embracing different ends. Czechs, Croats, Bosnians and Hungarians simply left their posts and went home. So when the Italians launched their offensive on 24 October, they were hardly attacking an army at all, just some pathetic remnants. Within forty-eight hours the end had come. By then the Hungarians had ordered their troops home, a Czechoslovak state had been proclaimed in Prague, and Croatia, Slovenia and Dalmatia had left the empire. Austria sued for peace. The offer went disregarded until the Italian army had advanced a sufficient distance to declare a great victory. On 4 November it came to a halt and the Austrian surrender was accepted. The war in Italy was over.

Signing the Armistice, 11 November 1918. Allied representatives receive German delegates in the railway carriage of Marshal Foch (standing, second from the right) in the Forest of Compiègne.

The Salonika front, virtually stationary since 1915, was also in the process of being transformed. A new commander (the distinguished and successful General Louis Franchet d'Esperey), and a considerable increase in heavy artillery, along with the effect of the German reverses on the Western Front, produced a new situation. A major Allied attack on 14 September precipitated a Bulgarian collapse. By the end of the month, an armistice had been granted and Allied troops fanned out towards the Danube.

The German commanders later claimed that it was the Bulgarian collapse which led them to seek an armistice. In fact the reverse was true. Small powers like Bulgaria hung on in the war as long as Germany had any prospect of success. Only when that hope disappeared did they leave the war with celerity. In this regard the war in Italy and the Balkans, like that in the east, was always contingent on the war in the west.

CONCLUSION

THE PEACE SETTLEMENT AND BEYOND

THE SIGNING OF THE Treaty of Versailles in the Hall of Mirrors on 28 June 1919: the famous painting by Sir William Orpen. In the centre of the painting are shown (from left to right) the three architects of the victory, David Lloyd George, Georges Clemenceau and Woodrow Wilson.

THE PEACE SETTLEMENT AND BEYOND

RECONSIDERATION

The peace settlement which followed the ending of the Great War has long been a subject of ill-repute. This requires explanation.

The settlement was devised by the victor states in the aftermath of a prolonged and bloody struggle which (not without reason) they saw as having been forced upon them. So it was likely to contain severe, and sometimes ill-advised, aspects. Yet the peace terms were not in the main malevolent; they contained much that was positively commendable, and anyway they were soon amended. That hardly provides grounds for regarding the peace settlement as fundamental in generating the international misfortunes which soon beset Europe.

Certainly, the peace treaty deprived Germany of some of its pre-1914 territory. But this was not evidence of a malicious determination to dismember Germany. Nor did it constitute a violation of President Wilson's much-praised Fourteen Points. Germany was obliged to part with conquered territories inhabited by mainly non-German peoples. These ranged from Polish lands annexed in the eighteenth century, through territory seized from Denmark and France in wars between 1864 and 1871, to the decidedly predatory conquests of 1914–18 made at the expense of Belgium and France and (in particular) Russia. All this was entirely in line with the Fourteen Points.

What the terms of peace did not do was seek to rob Germany of its territorial integrity, its unity, its imposing industries or its potentiality for great power status. In sharp contrast to what occurred in 1945, and for four decades thereafter, there was no enforced division of Germany into distinct states under various forms of external control. Even that strategically placed segment of Germany, the Rhineland, which the French military – for reasons of their nation's security – wanted to see converted into a French dependency, was left as an integral part of the German state.

Nor was any attempt made to demolish or bring under Allied control the heavy-industrial complex which had provided the matériel basis for the German war machine. That is, notwithstanding the oft-made assertion that the peace settlement embodied a determination to keep Germany henceforth enfeebled and dependent, its terms fail to reveal this. (The far greater severity of 1945 – if not for long thereafter – has occasioned no similar breast-beating.) Germany emerged from the war of 1914–18 as still the principal nation and mightiest industrial power on the continent of Europe. So it retained, in population and economic muscle (if not in actual weaponry), the wherewithal to wage another war. No statesman among the victor powers strove to make it otherwise.

If, nevertheless, the peace settlement sought to prevent Germany from again delivering the pre-emptive strike westward of August 1914, it was by means other

THE PEACE SETTLEMENTS IN EUROPE

Europe after the war, indicating the main changes in boundaries. Despite later German protestations, the map clearly shows how little territory Germany lost.

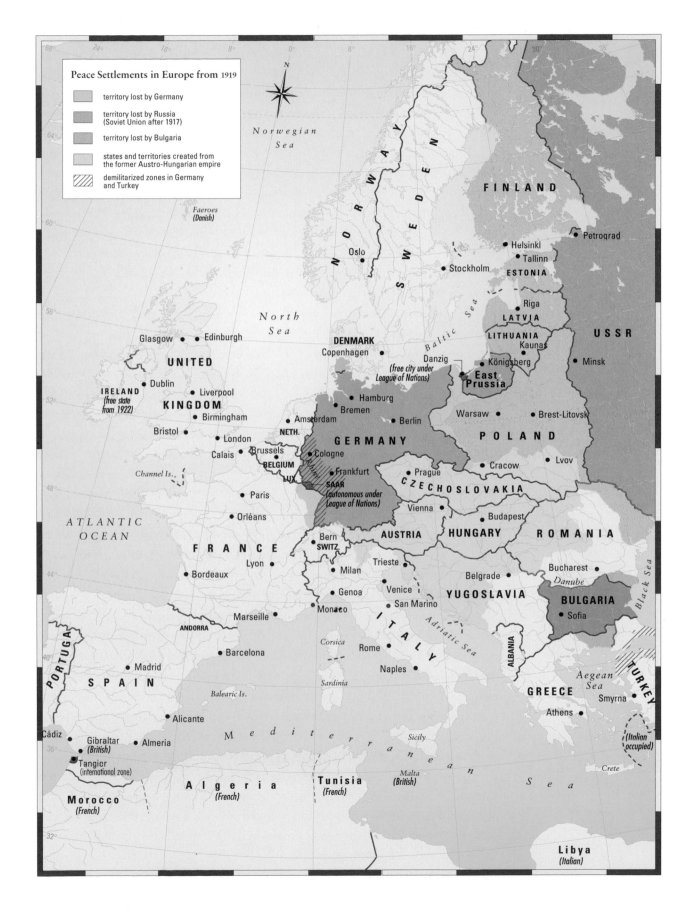

Peace Settlements in Europe from 1919

territory lost by Germany

territory lost by Russia
(Soviet Union after 1917)

territory lost by Bulgaria

states and territories created from
the former Austro-Hungarian empire

demilitarized zones in Germany
and Turkey

than Germany's dismemberment or substantial diminution. Germany was forbidden to maintain a large conscript army, or to possess major offensive weapons such as battleships and submarines and a military airforce. It was denied the right to station military forces on the west bank of the Rhine. And, to ensure this state of demobilization in the Rhineland, it was required to accept for fifteen years the presence there of an Allied army of occupation.

These provisions, far from seeming a manifestation of excessive severity, appeared to the French to fall well short of providing them with an ongoing guarantee against renewed aggression. Germany's disarmed state was an uncertain quantity. It depended either on the willingness of successive German governments to abide by the provisions of the treaty, or on the persisting determination of the Allies to take action if they did not. Yet the obvious alternative – the separation of the Rhineland from Germany – was simply unacceptable to the British and Americans.

To set at rest French anxieties, the President of the United States opted to offer France a binding guarantee of its territorial integrity against external aggression. That is, whoever ruled in Germany, be they peaceful or revanchist, would know that an act of war against France would constitute war also against the United States (and Britain).

The Great War could hardly have brought forth a more convincing act of statesmanship, or a better hope for a peaceful Europe. In its light, the persisting denigration of the actions of the peacemakers in 1919 seems plainly unsustainable.

THE ISSUE OF COMPENSATION

Certainly, dismissal of the peace settlement usually centres on other (inherently less weighty) aspects of the treaty: in particular, those which required Germany to pay reparations for the damage inflicted on the countries against which it had waged war – France and Belgium most of all.

The war, it needs to be recalled, had been an anomalous proceeding. Although at the last Germany had emerged the defeated power, at almost no stage had conflict raged upon its territory. By contrast, from first day to last it had raged upon the soil of France and Belgium. In addition to men killed and maimed in these countries, and the wives and children left bereft of a breadwinner, huge tracts of their territory had been devastated, both by battle and by deliberate acts of despoliation carried out by conquering and retreating German forces. As long as the damaged soldiers and deprived families of the Allied countries must be supported by the state in the aftermath of war, and as long as the devastated regions of France and Belgium must be restored to their former condition, it seemed evident to the victors (and was not plainly unwarranted from a more detached position) that the nation which had initiated conflict and imposed damage should bear the burden.

The matter of feasibility, certainly, placed a large query upon any such proceeding. Nor were the peacemakers wholly unaware of this, despite manifestations of mindless revanchism which initially gripped some of them (for example, in Britain's lamentable 'coupon' election of November 1918). The actual determination of the sum Germany should be required to pay was delayed until after the post-war ferocity had in some measure abated. And the sum was curtailed by being confined to civilian damage (though on a pretty wide-ranging definition of civilian) and, more significantly, by being limited to what was deemed Germany's capacity to pay.

Whether, even so, the amount eventually decided upon fell, in economic terms, within the bounds of the practicable remains a matter of dispute. What needs to be stressed is that the amount demanded was not, as seemed to some, intended to wreck the German economy. For any such intent would have rendered unobtainable the reparations which the damaged states hoped to secure. Also on the matter of practicality, it should be noted that the devastated regions of France and Belgium were indeed restored, and their maimed men, war widows and orphaned children were indeed supported, except that the economic burden of doing this had to be borne by the injured states. There is a further matter deserving of reflection. Between 1940 and 1944 France paid huge reparations (called 'occupation costs') to its conquerors. Confronted with masters as ruthless as the German Nazis, there was no point in trying to contest the economic feasibility of their demands. Germany after 1919 was not dealing with victor powers of that sort.

In the end, the attempt to levy reparations after the First World War foundered not on the question of whether either the initial sums demanded, or the diminished sums which soon replaced them were realistic or wildly inflated. The attempt failed because, first, the governments and people of Germany never accepted that such payments should be required of them and, second, the victor powers (after a brief show of determination by France) never possessed the ferocity to discover whether such sums were capable of extraction.

DISAPPOINTMENTS AND ACCOMPLISHMENTS

Quite evidently, we shall not explain the unrelenting disfavour with which the peace settlement of 1919 has been regarded by juxtaposing its meritorious against its more questionable aspects and asking where the balance rests. The explanation lies elsewhere. Most of all it lies in the expectations which had developed concerning the outcome of the struggle, during and immediately after the war.

During the course of the war, large hopes for betterment had been aroused: that this would be a war which would abolish war, that it would render the world safe for democracy, and that it would usher in more just societies and lands fit for heroes. The aftermath proved otherwise. Far from democracy prevailing universally, the fall of crowned autocrats was too often succeeded, immediately or before long, by the establishment of dictatorships and tyrannies without

Victory celebrations on Bastille Day, 14 July 1919. President Raymond Poincaré with Marshals Joffre and Foch, among other dignitaries at the Hôtel de Ville, Paris.

precedent in remembered history. Far from the sacrifices of 1914–18 ushering in a more secure world economically, in which the masses would enjoy regular employment, adequate housing and a sufficient health service, the post-war world witnessed a different scenario. During the 1920s, economic downturn and mass unemployment befell Britain, financial instability overtook France, and runaway inflation emerged in Germany. From 1929 these were followed by the onset of a vast depression blanketing the whole world. As for the ending of wars,

within twenty years of the peace settlement, first continental Europe and then the great powers of Asia and North America were plunged into a conflict of such magnitude and human cost as surpassed anything experienced between 1914 and 1918.

These melancholy events seemed to fly in the face of what were deemed the opportunities created by the Great War. The forces which had initiated conflict — in particular, authoritarian governments and hereditary rulers worshipping military values – had gone down in ruin. Liberal ideals and representative government had triumphed, even in those matters where autocracy and militarism were supposed to excel: the mobilization of resources for war purposes, and the appropriate employment of weaponry and military tactics upon the battlefield. Seemingly, therefore, the leaders of the liberal democracies meeting in Paris in 1919 were in a position to make all things new. President Woodrow Wilson, in his person and in his Fourteen Points, embodied the hopes for, and the opportunity to bring forth, a more benevolent world. If what emerged was not benevolence but tyrannies and economic adversity and a relentless march towards yet more terrible conflict, the fault apparently lay with the peacemakers.

In a widely endorsed view, the high-minded but vain and shallow Woodrow Wilson speedily fell victim to the more skilful and unscrupulous leaders of France and Britain – the vengeful Georges Clemenceau and the duplicitous Lloyd George. Their determination and chicanery, allied with Wilson's lack of substance, resulted in a peace settlement at odds with the idealism of the Fourteen Points and in violation of the principles of justice. So the way was opened to a terrible denouement.

This is curious. Much that might be considered statesmanlike, and that was certainly in conformity with the Fourteen Points, found its way into the peace settlement. It is worth elucidating some of Wilson's points. They laid down that a state of Poland, with access to the sea, should be restored to the map of Europe; that the former subject peoples of Austria–Hungary (such as Czechs and Slovaks) should be accorded autonomous development; that Belgium must be evacuated and restored; that the territory of France overrun by Germany should be relinquished, and 'the wrong' done to France in 1871 in the matter of Alsace–Lorraine 'should be righted'; that the lands of Russia conquered and annexed by Germany in 1917–18 should be evacuated and its territorial independence restored; and that a general association of nations (that is, a League of Nations) should be instituted 'for the purpose of affording mutual guarantees of political

Georges Clemenceau, Prime Minister of France in the last year of the war: 'Toujours, je fais la guerre.'

independence and territorial integrity'. Far from Wilson being 'bamboozled' (in J. M. Keynes's oft-quoted and utterly tendentious expression) by Clemenceau and Lloyd George into abandoning his ideals, all these meritorious aspects of the Fourteen Points were set in place by the peacemakers of Paris.

All these provisions, it also needs to be noticed, were in the forefront of Europe's plunge into renewed conflict between 1933 and 1941. In the course of these years, the rulers of Germany repudiated the League of Nations, remilitarized the Rhineland, terminated by threat or by military aggression the independence of Czechoslovakia and Poland, overran Belgium and France, and launched a massive invasion of the Soviet Union. To all appearances, therefore, it was not the supposed grotesqueries of the peace settlement, or any violations it might have contained of Wilsonian idealism, that brought Europe to war in the following decades. It was the peace settlement's devoted embodiment of Wilson's noble aspirations.

THE FAILURE OF ENFORCEMENT

As a rule, adverse judgements on the peace settlement stem from a delusory premise. Europe in 1919 was not a clean slate, and the peacemakers were not in a position to make all things (or indeed many things) new. While some of the states which had gone down to defeat would not again influence the course of events, the most important among them, Germany and Russia, most certainly would. There was little the peacemakers could have done to decide their subsequent actions. As for the successor states which had emerged from the Austro-Hungarian empire, they had established themselves in advance of the peace conference and would be only marginally influenced by its decisions. And of even greater import, the mighty coalition of powers which had accomplished the victory of 1918, and alone could ensure the endurance of any terms of settlement, was anything but stable and ongoing.

In short, neither the military accomplishments of 1918 nor the peace treaty of 1919 possessed the assurance of permanence, not because they lacked merit but because they were soon deprived of the power-base necessary to ensure their persistence. The forces in the major defeated power which had contributed so signally to generating a war, and which had sought to secure from that conflict not only territory but European hegemony, had been neither abolished nor rendered impotent by defeat on the battlefield. And the coalition of victor states which had accomplished Germany's defeat soon ceased to possess the resolve within and the unity of purpose among themselves which were absolutely essential to the preservation of their military triumph and of the terms of settlement they had put in place.

Germany and its allies had been overthrown by a combination of opponents which at various stages had included France, Russia, Britain, Italy and the United States. That combination had already lost one of its members before the close of the conflict, with the collapse of the tsarist regime and the disintegration of its liberal successor. The forces which seized power in St Petersburg late in 1917 and gradually extended their control to the whole of the former tsarist empire were fundamentally hostile to the liberal democracy of the West and eager to undermine the settlement devised in Paris. For this purpose, at crucial stages between 1919 and 1941 the rulers of Russia were prepared to assist those forces within Germany determined to subvert and finally overthrow the essential provisions of the peace settlement: those which limited Germany's military capacity to a level rendering it incapable of renewed aggression, and those which reinstated Poland as an independent power in the centre of Europe. The Soviet Union's actions during the 1920s in clandestinely encouraging German rearmament, and in 1939 in facilitating Poland's second obliteration, were key events in the elimination not just of the meritorious features of the peace settlement but of peace itself.

Russia's actions were not singular. Powers which had persisted in the war and contributed to making the peace soon ceased to be enforcers of the settlement of 1919. Italy, having entered the war for gain and paid dearly in the outcome, emerged from the peace conference with a strong sense of grievance and with its system of government profoundly undermined. Steadily thereafter it became the endorser and ally of international gangsterism.

Most significant of all, in 1920 the United States, whose entry into the war three years earlier had more than offset the withdrawal of Russia and rendered certain the victory of the Allies, terminated abruptly its participation in the maintenance of world order. This was the key event in emasculating the League of Nations as an arbitrator and law-enforcer between nations, in undermining enforcement of the peace settlement, and in presenting the defeated powers with the opportunity to reverse the military verdict of 1918. That is, there was nothing in the terms of the peace settlement – be those terms wise or foolish, high-minded or vindictive – that doomed Europe to instability, intimidation and another bloodbath. What did so was the repudiation by this key player of the peace settlement to which its president had contributed so signally, and often so wisely.

The rejection by the United States Senate of both the peace treaties and the guarantee to France, following so closely on Russia's defection into potential enmity towards the West, shattered the balance of power in Europe. In 1914 the rulers of Germany had confronted the combined might of France, Russia and Britain, and in 1917 of France, Britain and the USA. After 1919 the most they had to reckon with was half-hearted resistance to treaty violations by Britain and France. Such resistance could not long be maintained. Britain was grappling to come to terms with the human and economic cost of war. France was shaken to its foundations both by the scale of its wartime casualties – a scale proportionately well in excess of anything endured by its enemies – and by the insubstantial nature of the security which had been purchased by its prodigious endeavours and sacrifices. So, in the event, the forms of curtailment set in place to prevent Germany from again becoming a potential international menace were deprived of substance and gradually fell away one by one.

In the absence of the United States and Russia, Britain steadily drew back from the role of peace enforcer, hoping to reclaim its pre-war standing and economic confidence by unilateral resort to low armaments and international benevolence. As for France, it simply could not enforce unaided a victory which on its own it had plainly been unable to accomplish on the battlefield. The moment of truth came in 1923, when the French attempt to exact reparations by force through occupation of the Ruhr ended by undermining the franc. In the outcome the French government was forced to accept a rescue package which among other provisions required it to surrender the determination of reparations. Thereafter, France sought to preserve itself by arranging alliances with minor states on the far side of Germany, encouraging (along with Britain) such guarantees of borders with Germany as would dissuade the Germans from again disrupting the peace, and erecting a concrete wall which – in the worst of all possible worlds – would keep them out.

As a consequence, just in the ten years following the peace settlement, reparations were repeatedly scaled down, violations of the disarmament clauses of the treaty by Germany went disregarded, and Germany was welcomed into full membership of the League of Nations. Peace, hereafter, would be maintained by a presumed revulsion against war, a hoped-for goodwill among peoples, and an anticipated reasonableness among rulers. Appropriately, the culminating event of the 1920s was a major act of appeasement: the withdrawal, ahead of time, of the last Allied occupation troops from the Rhineland.

What followed immediately was noteworthy. In the German elections of 1930, Adolf Hitler's revanchist Nazi Party, full of denial that the German army had ever lost the war and bent on reversing the military verdict of 1918 (which the Nazis attributed to the stab in the back of the army by Jews, democrats and communists), was elevated from the status of an insignificant rump to that of a major force in German politics. Such a transformation, occurring at the moment when the world was starting to slide into deepening economic depression, was indeed ominous. It appeared to run counter to any notion that conciliation and concession had more to offer the cause of international stability than a policy of strict treaty enforcement.

FINAL THOUGHTS

The failure of the Great War to produce its expected rewards – lasting peace, an ongoing prosperity, and the ever-widening acceptance of democracy – has caused that struggle to be viewed negatively. When commentators in present times wish to discourse on the futility of war, they usually seize upon the conflict of 1914–18 to provide historical endorsement.

This judgement requires, and is usually not given, serious scrutiny. The war had certainly shown that, with the application of industrialization to the business of battle, armed conflict had become – even more than in the past – a cruelly destructive transaction, which wasted treasure and swept away millions of fit young males. In these powerful respects, war was (as ever) a denial of aspirations towards human betterment.

But this is not all that should be said. Offsetting, if only in a measure, warranted revulsion against this war – for its appalling human cost, its contribution to economic dislocation, and the emergence of yet more terrible forms of tyranny – are three elements.

First, particularly in Britain, war helped to generate, if haltingly, notions of a more equal and a more benign society. Women received the vote, the working week was reduced to a more tolerable level, and people thrown out of work acquired some entitlement to unemployment benefits. This was a far cry from the land fit for heroes which had been promised. But it pointed in a direction which later generations would embrace.

Second, the war advanced the notion that identifiable nationalities were entitled to self-government and self-determination. The application of this principle, certainly, did not always yield admirable consequences. Yet the termination of alien rule in parts of eastern Europe and Arabia plainly helped to lessen the injustices of the world.

But it is the third aspect that most requires emphasis. Manifestly, in August 1914 the status quo of western Europe was about to vanish. Either the liberal democracies would engage in a terrible episode of bloodletting in order to preserve their independence, territorial integrity and great power status, or they would avoid bloodshed by permitting the autocracy and militarism of the kaiser's Germany to overwhelm them. That is, the alternative to the horrors of this war was not the continuation of the existing order. It was western Europe's abandonment of some of its finest achievements. These achievements derived from its struggles against absolute church and absolute monarchy, and from its endorsement of the principles of the Enlightenment: elected governments, freedom of speech and of conscience, respect for the rights of minorities, and at least partial acknowledgement of the notion that all people are created equal and possess the same entitlements to life, liberty and the pursuit of happiness.

Entente Cordiale. Premier Clemenceau with Sir Douglas Haig (British commander-in-chief), Sir Julian Byng (commander of the British Third Army), and French officers.

The menace presented to these values by militant, expansionist kaiserism may not have equalled the horrors presented twenty-five years later by aggressive Nazism. But it constituted horror enough: the reversal of so much that had been achieved towards human betterment in the centuries just past, and the desolation of hopes for yet further advancement. Correspondingly, the Great War's vindication of liberal ideals and democratic forms, and its severe rebuff to rampant authoritarianism, was a mighty accomplishment.

At the present time, when – anyway in the West – the decencies and humanity inherent in the liberal state are so widely experienced and appreciated, it is hardly appropriate for Westerners to denigrate a bygone struggle which served to vindicate those values and preserve those decencies from deadly peril.

BIOGRAPHICAL DETAILS

ALLENBY, MAJOR GENERAL SIR EDMUND (1861–1936)
On the Western Front he commanded the Cavalry Division 1914, the Cavalry Corps 1915, and Third Army 1916–17. He was appointed commander-in-chief Palestine in 1917–18. One of the few generals to emerge from the war with an enhanced reputation, he was perhaps fortunate to be transferred to a theatre where he could engage a declining enemy with superior resources. Nevertheless he conducted the final campaign against the Turks with skill and determination.

ASQUITH, HERBERT (1852–1928)
British Prime Minister 1908–16. Asquith's skilful handling of the crisis of late July–early August 1914 ensured that he led his Liberal government united into war. As war leader he suffered from a lack of knowledge about or interest in military affairs and the difficulties of improvising a mass army from scratch. A coalition formed with the Conservatives in May 1915 was not a success. He was replaced by the more dynamic Lloyd George in December 1916 and did not again hold office.

BIRDWOOD, LIEUTENANT GENERAL SIR WILLIAM (1865–1951)
Commander of the Anzac Forces in Gallipoli 1915, in the Western Front 1916–17, I Anzac Corps 1917–18, Fifth Army 1918. At Gallipoli his tactical handling of the Anzac forces made the best of an impossible undertaking. On the Western Front his career as corps commander demonstrated the limited initiative that could be exercised at that level of command. In 1918 his army efficiently followed up the retreating Germans.

BRUSILOV, GENERAL ALEXEI (1853–1926)
Commander of the Russian Eighth Army 1914–15, South-Eastern Front 1916–17, and commander-in-chief 1917. He was an energetic commander who launched his country's only successful offensive on the Eastern Front. In 1916 he attacked on a 300-mile front and pushed back Austro-Hungarian forces some 50 miles before German forces arrived to stabilize the situation. In 1917 he was given the thankless task of carrying out Kerensky's offensive with a disintegrating army. He was replaced by Kornilov

in July. He ended his career as a general in the Red Army.

CADORNA, GENERAL COUNT LUIGI (1850–1928)
An artilleryman, and chief of the Italian general staff June 1914–November 1917. A severe disciplinarian, he led an under-equipped army in a succession of bloody and futile assaults against Austro-Hungarian forces along the River Isonzo. He was sidelined after the disastrous Italian reverse at Caporetto in November 1917.

CHURCHILL, WINSTON (1874–1965)
First Lord of the Admiralty 1914–15, Chancellor of the Duchy of Lancaster 1915, soldier 1916, Minister of Munitions 1917–18. His desire to use the navy to influence the war on land led to the poorly thought out Dardanelles campaign. Its failure resulted in his removal from high office, although he later secured a partial restoration with his appointment as Minister of Munitions. He continued to pour out strategic and tactical ideas (some of them good) until the end of the war. His memoir, *The World Crisis*, though partial, is one of the classics of war literature.

CLEMENCEAU, GEORGES (1841–1920)
Member of the French Senate 1914–17, Premier 1917–18. He became Premier due to the setbacks of 1917, his unrelenting criticism of a succession of French governments and the growth of a widespread belief that the war should be prosecuted more vigorously. He remained resolute in the face of the German onslaught in the West in 1918, was instrumental in maintaining unity with the British and securing the Supreme Command for Foch. In 1919 he was a leading figure at the peace conference where his reasonable attempts to ensure security for his country were soon criticized outside France as harsh and inflexible.

FALKENHAYN, GENERAL ERICH VON (1861–1922)
Minister of War 1914, chief of the general staff 1914–16, commander against Romania 1916–17 and in Palestine 1917–18. A convinced westerner, his offensives in the west brought fearsome losses upon the German army at Ypres and Verdun. These failures and

the mounting material superiority of the British army during the course of the Somme campaign, along with Romania's entry into the war against Germany, led to his supersession by Hindenburg and Ludendorff in August 1916. His armies overran Romania in nine weeks, but his failure to halt Allenby in Palestine saw him replaced by Liman von Sanders.

FOCH, GENERAL FERDINAND (1851–1920)
Commander XX Corps and Ninth Army in 1914, Northern Army Group in 1915–16, French forces at the Somme in 1916, chief of staff to General Pétain in 1917, generalissimo Allied forces on the Western Front in 1918. In 1918 his appointment as Supreme Commander was an important symbol of Allied determination to fight on in the face of the Ludendorff offensive. He presided over rather than directed the victory campaign in the second half of 1918.

FRENCH, FIELD MARSHAL SIR JOHN (1852–1925)
Commanded BEF 1914–15, Home Forces 1915–18. Out of his depth from the first British battle at Mons, he proved no strategist or tactician. Unlucky in being required by the exigencies of coalition warfare to participate in battles of which he disapproved, he was nevertheless temperamentally unsuited for the rigours and horrors of modern industrial war. He was replaced by Haig in December 1916.

GREY, SIR EDWARD (1862–1933)
British Foreign Secretary 1905–16. He was a major force in confirming the realignment of Britain with France and Russian in the years 1905–14 in order to counter the threat from Germany. In this he showed much skill, effecting the shift without compromising Britain's ability to act independently. His efforts to resolve the international crisis of July–August 1914 failed because of German determination for a continental war. In wartime he was less effective. Failing eyesight and his lack of enthusiasm for Lloyd George saw him leave the government in December 1916.

HAIG, GENERAL (LATER FIELD MARSHAL) SIR DOUGLAS (1861–1928)
Commanded I Corps in 1914, First Army in 1914–15, commander-in-chief BEF in 1915–18. He is often characterized as a dullard whose only strategic or tactical insight was to wear down the enemy forces

at an equivalent cost to his own. In fact his campaigns never aimed at attrition. They were designed to end the war by inflicting such heavy blows as to produce a breakthrough. The designs were far too ambitious for the resources at his disposal. He seems destined to be remembered for the failures of 1916 and 1917 rather than the victories of 1918.

HAMILTON, GENERAL SIR IAN (1853-1947)
Hamilton had served as Kitchener's staff officer in the South African war and was brought out of retirement by his old chief to take command at Gallipoli. As commander, Hamilton was never given the resources to make a half-decent effort at what was an impossible task. Admired at the time as an intellectual general with a taste for poetry, the jolly insouciance with which he presided over the slaughter of his men has won him few admirers in recent times.

HINDENBURG, GENERAL (LATER FIELD MARSHAL) PAUL VON (1847–1934)
Brought from retirement in 1914 to command on the Eastern Front, Hindenburg was elevated to the position of chief of the general staff 1916–18. In the east he scored a series of spectacular if indecisive victories in 1914 and 1915. His tenure as chief of the general staff saw the military arm take control of most aspects of German life. He is generally thought to have been a figurehead dominated by Ludendorff.

HÖTZENDORFF, GENERAL FIELD MARSHAL CONRAD VON (1852–1925)
Chief of the Austrian general staff 1914–17. His reckless adventurism played some role in propelling his country into war in 1914. His armies, while demonstrating endurance and more cohesion than might have been anticipated, proved unable to defeat Serbia or to hold the line against Russia without German assistance. He was sacked as an impediment to peace when the Emperor Karl succeeded Franz Joseph in February 1917.

JOFFRE, GENERAL (LATER MARSHAL) JOSEPH JACQUES CÉSAIRE (1852–1931)
French commander-in-chief on the Western Front 1914–16. In 1914 his disastrous offensive policy cost his armies 300,000 men in six weeks, but this was soon offset by his defensive manoeuvring which resulted in the victory of the Marne. From 1915 to 1916 it was his policy to expel the Germans from French soil. The

unimaginative nature and heavy cost of these endeavours saw him replaced by General Robert Nivelle late in 1916.

KEMAL, MUSTAPHA (ATATÜRK) (1881–1938)

Commander 19th Turkish Division 1915, Caucasus front 1916–17, Seventh Army 1918. At Gallipoli in 1915 his leadership in April and August helped thwart the Allied offensives. He assisted in retrieving the situation against Russia in the Caucasus in 1916. Against the British in Palestine in 1918 he conducted a skilful retreat but was unable to halt his opponents' advance. In the post-war years he became President of the Turkish Republic (1924–38), repudiated the Treaty of Sèvres and saved the Turkish heartland from dismemberment by the Allies.

KERENSKY, ALEXANDER (1881–1970)

Social revolutionary. After the fall of the tsar in the revolution of March 1917 Kerensky became Minister of War and then leader of the Provisional Government. Determined to continue the war, he launched the disastrous offensive in July 1917 which caused the final disintegration of the Russian armies. By November his support had collapsed and he was easily brushed aside in the Bolshevik seizure of power. He went into exile in late 1917 where he remained for the rest of his life.

LLOYD GEORGE, DAVID (1863–1945)

Chancellor of the Exchequer 1905–15, Minister of Munitions 1915–16, Minister of War 1916, Prime Minister 1916–22. He soon established a reputation as the most vigorous member of Asquith's wartime government. His achievements as Minister of Munitions enhanced his standing. He replaced Asquith as Prime Minister in December 1916. In that role his organizational abilities on the Home Front were a vital factor in British mobilization. His relations with his generals were less satisfactory and his failure to impose an alternative policy on them during the Third Ypres offensive detracted from his wartime achievement. However, he provided firm leadership of the nation during the setbacks and successes of 1918.

LUDENDORFF, GENERAL ERICH VON (1865–1937)

Quartermaster-general Second Army 1914, chief of staff to Hindenburg 1914–16, first quartermaster-general 1916–18. He played a large role in the German victories in the east at Tannenburg, the Masurian

Lakes and into Russian Poland (1914–15). From 1916, despite his undistinguished title, he exercised effective control over most aspects of German life. His misguided western offensive in 1918, coincident with the aggressive expansionism he pursued in the east, helped bring Germany to ruin.

MACKENSEN, GENERAL AUGUST VON (1849–1944)

Commander of Eleventh Army in Galicia 1915 and German forces in Serbia 1915, army commander under Falkenhayn in Romania 1916, head of army of occupation in that country 1916–18. He carried out much of the detailed planning which brought Germany some of its most spectacular victories in the east: Gorlice–Tarnow and Serbia in 1915 and Romania in 1916.

MOLTKE, GENERAL HELMUTH VON (THE YOUNGER) (1848–1916)

Chief of the general staff 1914. Nephew of the victor of the Franco-Prussian War 1870–71, he had the unenviable task of converting the over-ambitious plan drafted by Schlieffen into some sort of military reality. His modifications were generally sensible but could not procure the plan's success. After the reverse at the Marne in September 1914 he was replaced by Falkenhayn.

NIVELLE, COLONEL (LATER GENERAL) ROBERT GEORGES (1856–1924)

Artilleryman. Commanded artillery brigade 1914, divisional and corps commander 1915, Second Army 1916, French commander-in-chief 1916–17, commander North Africa 1917–18. His small-scale victories at Verdun in late 1916, accomplished by artillery innovation, saw him replace Joffre as French commander in the west. His rash promises of a rapid victory in the spring of 1917 and the subsequent failure of his offensive reduced the French army to mutiny. His fall (he was replaced by Pétain in May 1917) was no less spectacular than his rise.

PÉTAIN, GENERAL PHILIPPE (1856-1951)

At the outbreak of the war Pétain held the humble rank of colonel and was in charge of a regiment. His abilities soon saw him rise to the command of a division, (1914) then an army (1915). His finest moments of the war came with his defence of Verdun in 1916 and in his rehabilitation of the French army after the Nivelle disaster in 1917. In 1918 his pessimism over Allied prospects after the Hindenburg offensive resulted in Foch being given the supreme command.

PLUMER, GENERAL SIR HERBERT (1857–1932)
Commanded II Corps 1914, Second Army 1915–17, Allied forces in Italy 1917–18, Second Army 1918. His reputation is closely linked to the Ypres salient which the Second Army held from 1915 to 1918. His finest moment came at the battle of Messines in June 1917 when nineteen huge mines destroyed the German positions on the ridge. Later he conducted the only successful phase of the Third Battle of Ypres, defeating the Germans in three set-piece attacks, but then directed its abysmal last phase unprofitably. Events in 1918 saw his army reduced to a flanking force of the main advance. His last appointment was as commander of British occupation forces in Germany.

RAWLINSON, GENERAL SIR HENRY (1864–1925)
Commander 7th Division 1914, IV Corps in 1914–15, Fourth Army in 1916–18. He was one of the first commanders on the Western Front to note the importance of artillery, and the futility of attempting more than tentative advances ('bite and hold'). His subservience to Haig and a lack of mental rigour prevented him from implementing these policies until 1918. So he presided over the greatest British disaster of the war (the Somme 1916) and the greatest British success (the Advance to Victory 1918).

ROBERTSON, MAJOR GENERAL (LATER FIELD MARSHAL) WILLIAM (1859–1933)
Quartermaster-general BEF 1914, chief of the general staff 1915, chief of the Imperial general staff 1915–18. He proved his worth during the British retreat from Mons, where his organizational ability did much to save the BEF. His record as CIGS was more ambiguous. He rightly identified the Western Front as the vital theatre, but he acquiesced in Haig's attempts at breakthrough at the Somme in 1916 and at Third Ypres in 1917. He was manoeuvred out of office by Lloyd George in February 1918 and replaced by the lightweight Sir Henry Wilson.

SCHLIEFFEN, GENERAL FIELD MARSHAL ALFRED VON (1833–1913)
As chief of the German general staff from 1888 to 1906 Schlieffen drew up the plan with which the Germans launched the invasion of Belgium and France in 1914. At times Schlieffen acknowledged the impracticality of his plan but carried out little logistical calculation to give substance to his plans or his fears.

WILHELM II, KAISER (1859–1941)
A weak leader who concealed his vacillating nature with frequent bellicose pronouncements. His scheme to build a battle fleet was the decisive factor in alienating Britain before the war and his erratic diplomacy in these years did much to cement the Anglo-Franco-Russian Entente. During the war his opinions were frequently disregarded by the military. Under the regime of Hindenburg and Ludendorff he was reduced to a minor role. He abdicated on 9 November 1918 and spent the remainder of his life in exile in Holland.

WILSON, WOODROW (1856–1924)
President of the United States 1912–20. In 1914 and in the election campaign of 1916 he determined to maintain American neutrality. However, his liberal sentiments, Germany's employment of submarine warfare against neutral as well as Allied merchant ships, and clumsy diplomacy by the German foreign office caused the USA to enter the war in April 1917. His Fourteen Points (January 1918) provided the basis for the Armistice. At the peace conference in 1919 his main objective was to secure self-determination for suppressed nationalities in eastern Europe and the acceptance of the League of Nations. His plans and his health collapsed when the Senate refused to ratify the Treaty of Versailles and he died an invalid.

FURTHER READING

The volume of books dealing with the Great War of 1914–18, in all its manifestations, is vast. Their value, of course, varies hugely, but the number possessing sufficient merit to deserve entry in a select bibliography remains considerable. It certainly exceeds the space available here. So all that can be offered is a selection of some fifty works. Few among them are without shortcomings, but they usually possess distinct merits and will provide starting points for further investigation.

Possibly the most difficult aspect is itemizing general histories that are both wide-ranging and mainly satisfying. The most widely known, and certainly most readable, one-volume history, A. J. P. Taylor's *The First World War: an Illustrated History* (1963), is best considered a form of diversion and a display of dogma. By contrast, C. Falls, *The First World War* (1960), is solid and scholarly, but it is confined to military aspects and is not a lively read. Gerd Hardach, *The First World War* (1977), concentrates on economic aspects.

An extended work that demands attention, despite the difficulty of passing a simple judgement on it, is Purnell's *History of the First World War*, published in 128 weekly parts between 1969 and 1971 and subsequently reissued as the Marshall Cavendish *Encyclopedia of the First World War* (1984). Its articles are of variable quality, but it contains some, such as J. Keegan on trench war, which are first class, and there are few aspects of the war which it does not touch upon.

The outbreak of the war has itself been the subject of a large library. A solid and scholarly account is provided by L. Albertini, *The Origins of the War of 1914*, 3 vols. (1952–7). Good supplements are provided by R. Evans and H. P. von Strandmann, *The Coming of the First World War* (1988), and J. Joll, *The Origins of the First World War* (1992). And three books dealing with the role of particular countries in the outbreak of war cannot be passed over: Fritz Fischer, *Germany's Aims in the First World War* (1967) and *War of Illusions* (1975), and Zara Steiner, *Britain and the Origins of the First World War* (1977).

The course of battle on the various fronts has been thoroughly covered in some instances and little attended to in others. Best served, not surprisingly, is the conflict on the Western Front. The multi-volume *British Official History* by Sir J. Edmonds *et al.* is hard work, but generally sound and ultimately inescapable. The vexed issue of command, as it concerned the British Expeditionary Force, is dealt with from rather different viewpoints by J. Terraine, *Douglas Haig: the Educated Soldier* (1963), and R. Prior and T. Wilson, *Command on the Western Front* (1992). The major operations in France and Flanders are dealt with, generally satisfactorily, in many works. For 1914 and 1915 there is G. Ritter, *The Schlieffen Plan* (1959); J. Keegan, *Opening Moves* (1971); J. Terraine, *Mons* (1960); and R. Holmes, *The Little Field-Marshal: Sir John French* (1981). For 1916 and 1917 see Alistair Horne, *The Price of Glory: Verdun 1916* (1962); A. Farrar-Hockley, *The Somme* (1964); E. L. Spears, *Prelude to Victory* (1939) (on the Nivelle offensive); and R. Prior and T. Wilson, *Passchendaele* (1996). For 1918, J. Terraine, *To Win a War* (1978), is a good starting point.

The conflicts on the Eastern and South-Eastern Fronts have not been covered with the same devotion. For the former, N. Stone, *The Eastern Front* (1975), and G. Jukes, *Carpathian Disaster* (1971), are of major importance. Much of value is also to be gleaned from G. E. Rothenburg, *The Army of Francis Joseph* (1976), and D. Showalter, *Tannenberg, Clash of Empires* (1991). For the South-Eastern Fronts, some important aspects are dealt with in C. Falls, *Military Operations, Macedonia*, 2 vols. (1933–5), and A. Palmer, *The Gardeners of Salonika* (1965).

The war on the Italian Front is ill-served, especially by books in English, but a start may be made with C. Falls, *The Battle of Caporetto* (1966).

Turkey's involvement in the Great War has attracted rather more attention, on account of the somewhat exotic locales and the high level of British involvement. Among the many books on the Gallipoli operation, the best place to begin is still R. R. James, *Gallipoli* (1965). For the war in the Western desert, a useful starting point is A. P. Wavell, *Allenby: a Study in Greatness* (1940). For the campaign in Mesopotamia, A. J. Barker, *The Neglected War* (1967), is almost alone in the field.

In addition to works on military episodes, there is a rich literature in English about at least some of the warring countries. Perhaps not surprisingly, Germany has been much written about, whereas Austria–Hungary is rather neglected. Two recent works of particular merit are H. Herwig, *The First World War: Germany and Austria–Hungary* (1997), and R. Chickering, *Imperial Germany and the Great War* (1998). Also of value for particular aspects are G. F. Feldman, *Army, Industry and Labor in Germany* (1966); J. Kocha, *Facing Total War* (1984); and M. Cornwall, *The Last Years of Austria–Hungary* (1990).

Britain during the Great War has been dealt with comprehensively in T. Wilson, *The Myriad Faces of War* (1986), and J. M. Bourne, *Britain and the Great War* (1989). Social aspects are the subject of powerful investigation in J. M. Winter, *The Great War and the British People*; the raising of the 'new' armies is covered in P. Simkins, *Kitchener's Army* (1988).

Authoritative works on France at war are not numerous. Attention should be given to the writings of J. J. Becker, including *The Great War and the French People* (1985). Also of value is P. Fridenson, *The French Home Front* (1992). For Russia, note should be taken of W. B. Lincoln, *Passage Through Armageddon* (1986), and, regarding a crucially revealing episode, J. W. Wheeler-Bennett, *Brest-Litovsk: the Forgotten Peace* (1938). For works on the USA's involvement in the Great War, a good starting place is E. Coffman, *The War to End All Wars* (1968).

The peace settlement was the subject of denunciation from the moment of its signature, in J. M. Keynes, *The Economic Consequences of the Peace* (1919). It has gone on being denounced ever since, notwithstanding the effective rejoinder to Keynes as long ago as 1940 by Etienne Mantoux, *The Carthaginian Peace: the Economic Consequences of Mr Keynes*, which happily has been reissued (1999). Other works on the peace settlement deserving attention include S. Marks, *The Illusion of Peace* (1976), and A. Sharp, *The Versailles Settlement* (1991).

The foregoing, it needs to be repeated, is no more than a rather arbitrary selection. The authors of this book are painfully aware of the many works of merit which the dictates of space have required them to omit.

INDEX

PICTURE ACKNOWLEDGEMENTS

Every effort has been made to contact the copyright holders for images reproduced in this book. The publishers would welcome any errors or omissions being brought to their attention.

A.K.G endpapers and pp 6, 14, 19, 22, 23, 25, 26, 27, 29, 32, 34, 39, 40–41, 43, 44, 47, 49, 50, 52, 56, 57, 58, 59, 60, 64–5, 66, 67 top, 68, 70, 73, 75, 78, 79, 80–81, 83, 85 right, 94, 99, 101, 105, 110, 113, 115, 116, 133, 135, 136, 138–9, 139, 142, 152–3, 155, 156, 157, 158, 163, 164–5, 166, 167, 168–9, 172, 180, 199, 200, 206–7, 208; Art Archive pp. 48–9, 85 top, 120, 130, 132, 148–9, 150, 154, 162, 176 bottom, 196; Corbis-Bettmann/UPI pp. 30, 46, 62, 67 (bottom), 76, 86, 87, 95, 107, 126–7, 137, 170–71, 174–5, 179, 186–7, 194–5; Hulton Getty pp. 108–9; Imperial War Museum pp. 28 (Q.81554), 35 (Q.43033), 45 (Q.51501), 53 (Q.53759), 90 (Q.24168), 96 (Q.4123), 100 (Q.53749), 102 (Q.30014), 112 (Q.54534), 118–9 (Q.930), 119 (Q.5818), 123 (Q.1388), 125 (Q.7269), 129 (Q.2980), 134 (Q.2017), 141 (Q.5171), 144–5 (Q.3217), 146 (E.AUS 648), 147 (Q.2295), 160 (Q.9534), 175 (Q.10786), 176 top (Q.10816), 181 (Q.10857), 183 (Q.9269), 185 (Q.6948), 191 (Q.9382), 192–3 (Q.9364), 213 (C.0.3399); Peter Newark's Pictures pp. 33, 98, 106, 117, 122, 177, 190.

Drawings on the title page and on pages 2, 38, 124, 140 and 185 are by Peter Smith and Malcolm Swanston of Arcadia Editions Ltd

ENDPAPER: *British C Division attack at Achi Baba. Achi Baba, a 200-metre hill that dominated the Helles sector of the Gallipoli Peninsular, was the Allied commander-in-chief General Hamilton's first main objective following the Gallipoli landings on 25 April 1915.*